T0135527

Controller and Network Design Exploiting System Structure

Von der Fakultät Konstruktions-, Produktions- und Fahrzeugtechnik
der Universität Stuttgart
zur Erlangung der Würde eines Doktors der
Ingenieurwissenschaften (Dr.-Ing.) genehmigte Abhandlung

Vorgelegt von

Simone Schuler

aus Tübingen

Hauptberichter: Prof. Dr.-Ing. Frank Allgöwer
Mitberichter: Prof. Dr. Carsten W. Scherer
Prof. Daniel Zelazo, PhD

Tag der mündlichen Prüfung: 02. Mai 2014

Institut für Systemtheorie und Regelungstechnik der
Universität Stuttgart

2015

Bibliografische Information der Deutschen Nationalbibliothek

Die Deutsche Nationalbibliothek verzeichnet diese Publikation in der
Deutschen Nationalbibliografie; detaillierte bibliografische Daten sind
im Internet über http://dnb.d-nb.de abrufbar.

D 93

©Copyright Logos Verlag Berlin GmbH 2015
Alle Rechte vorbehalten.

ISBN 978-3-8325-3924-5

Logos Verlag Berlin GmbH
Comeniushof, Gubener Str. 47,
10243 Berlin
Tel.: +49 (0)30 42 85 10 90
Fax: +49 (0)30 42 85 10 92
INTERNET: http://www.logos-verlag.de

Für Dominik, Annika und Emil.

Acknowledgments

The results presented in this thesis are the outcome of my time as a research and teaching assistant at the Institute for Systems Theory and Automatic Control (IST) at the University of Stuttgart. There are several people to whom I would like to express my gratitude for their positive influence on this thesis.

First and foremost, I would like to thank my advisor Prof. Frank Allgöwer. He gave me lots of freedom in my research and created with his enthusiasm and broad knowledge an unique and open minded, internationally renowned environment at the institute. Moreover, he provided support for attending several international conferences, where I was able to present my work and discuss my findings with other researchers.

Second, I want to thank Prof. Daniel Zelazo for the very fruitful interaction, for sharing his ideas and directing me to new control problems. I would also like to thank Prof. Carsten Scherer for joining the committee of my PhD defense and for all the knowledge I gained from his lectures and short courses.

My time at the institute would not have been such a joyful and productive time without all the colleagues and guests. Thank you all for giving support or distraction as needed. I would also like to thank my students who supported my research and teaching assistance.

Most of all, I would like to thank my family for their support and patience over all those years.

Garching bei München, November 2014
Simone Schuler

Contents

Symbols and Acronyms

Symbols

$\mathbb{R}$	set of reals
$\mathbb{R}^+$ ($\mathbb{R}^-$)	set of positive (negative) reals
$\mathbb{R}_0^+$ ($\mathbb{R}_0^-$)	set of non-negative (non-positive) reals
$\mathbb{C}$	set of complex numbers
$\mathbb{C}^+$ ($\mathbb{C}^-$)	set of complex numbers with positive (negative) real part
$\mathbb{C}_0^+$ ($\mathbb{C}_0^-$)	set of complex numbers with non-negative (non-postive) real part
$\|\cdot\|_p$	p-norm of vectors and matrices, for $p = \{0, 1, 2, \infty\}$
$\|\cdot\|_{\ell_p}$	p-norm of signals, for $p = \{0, 1, 2, \infty\}$
$M^\perp$	orthogonal complement
$\mathrm{diag}(a_i)$	matrix with elements a_i on the (block-) diagonal
I	identity matrix of appropriate dimension
0	zero matrix of appropriate dimension
$\mathbb{1}$	vector of all ones of appropriate dimension
$A \circ B$	Hardamard product of A and B
$\langle \cdot, \cdot \rangle$	standard inner product between two vectors
$\{x(k)\}$	infinite dimensional sequence $x = \{x(1), x(2), \dots\}$
$f * g$	convolution of two discrete time signals f and g
$\mathcal{F}_l(P, K)$	lower fractional transformation of P and K
$\lambda(A)$	eigenvalue of matrix A
$\mathrm{Tr}[A]$	trace of matrix A
$\mathrm{vec}(A)$	vectorized matrix
$A > 0$ ($A \geq 0$)	positive (semi-) definite matrix
Σ	dynamical system
$\left[\begin{array}{c|c} A & B \\ \hline C & D \end{array}\right]$	transfer function $C(sI - A)^{-1}B + D$ of a dynamical system
$\mathcal{P}_N(x)$	truncation operator $\{x(1), x(2), \dots, x(N), 0, 0, \dots\}$
$\mathcal{G}$	graph
$\mathcal{V}$	node set of a graph
$\mathcal{E}$	edge set of a graph
$\mathcal{T}$	spanning tree
$\mathcal{K}_n$	complete graph on n nodes
E	incidence matrix of a graph
L	Laplacian of a graph
L_e	edge Laplacian of a graph
μ	algebraic connectivity of a graph, i.e. $\lambda_2(L)$

Acronyms

LTI	linear, time-invariant
RSN	relative sensing network
MAS	multi-agent system
LMI	linear matrix inequality
FIR	finite impulse response
IIR	infinite impulse response
SDP	semidefinite program
LP	linear program

Abstract

This work considers the problem of decentralized controller and network design under communication constraints for homogenous and heterogenous interconnected systems. Different from most existing approaches, where the topology of the controller or network is fixed *a-priory* and only the link dynamics are designed, we integrate topology and dynamics design into one overall optimization problem. Structure optimization is done subject to pre-defined performance constraints in terms of the $\mathcal{H}_2$ and $\mathcal{H}_\infty$-norm of the closed loop system. We develop computationally efficient formulations by convex relaxations which makes the proposed design methods attractive for practical applications. We further introduce the concept of an ℓ_0-system gain for discrete-time LTI systems. With this newly introduced system gain, we give a system theoretic explanation of the sparse closed loop response of ℓ_1-optimally controlled systems.

Deutsche Kurzfassung

Motivation und Forschungsbeiträge

Die Analyse und Regelung von vernetzten Systemen ist eine der großen Herausforderungen der modernen Ingenieurwissenschaften (siehe auch Murray, 2002). Anwendungsgebiete erstrecken sich von stark verkoppelten chemischen Anlagen, Wärmetauschern, chemischen Reaktionsnetzwerken und Energienetzen bis hin zu Kommunikationsnetzwerken wie dem Internet, Transportsystemen bestehend aus Kraftfahrzeugen, Zügen und Flugzeugen und Umgebungsüberwachung durch unbemannte Flugobjekte. Einerseits handelt es sich hierbei um Teilsysteme die autonom agieren wollen, aber durch Kopplungen mit anderen Teilsystemen beeinflusst und gestört werden, und andererseits um Gruppen von autonomen Agenten, die eine gemeinsame Aufgabe erfüllen sollen. Unabhängig vom Anwendungsgebiet setzen sich diese Systeme folgendermaen zusammen: aus der Dynamik der individuellen Teilsysteme, den Kopplungen zwischen den einzelnen Teilsysteme, sowie der Kopplung der Teilsysteme mit dem Regler und dessen Reglerarchitektur.

Mit zunehmender Verbreitung von vernetzten Systemen steigen auch die Anforderungen an die Regelgüte eben dieser. Zusätzlich zur Stabilität des Gesamtsystems werden oft Anforderungen wie beispielsweise schnelle Konvergenz bei Konsenzproblemen, Optimalität bezüglich einer Systemnorm oder Robustheit gegenüber Unsicherheiten gestellt. Um diesen gesteigerten Anforderungen Rechnung zu tragen, ist mehr Kommunikation zwischen den einzelnen Agenten notwendig und Kopplungen zwischen Teilsystemen können nicht länger ignoriert werden. Informationsaustausch ist im Allgemeinen jedoch teuer, da die einzelnen Systeme über ein großes Gebiet verteilt sein können (z.B. ganz Kontinentaleuropa bei Energienetzen) und daher begrenzt. Grundsätzlich erreicht ein zentraler Regler, der Zugang zu Informationen von allen Teilsystemen hat und diese auch beeinflussen kann, die beste Regelgüte. Wird der Informationsaustausch zwischen den Teilsystemen reduziert verschlechtert sich die erreichbare Regelgüte. Jedoch lässt sich mit geeigneter Wahl der verbleibenden Signale beinahe die Regelgüte des zentralen Reglers erreichen. Dabei stellt sich folgende Frage: Wie können die Signale innerhalb eines Netzwerkes identifiziert werden, die den größten Beitrag zu einer zufrieden stellenden Regelgüte leisten? Diese Frage kann auf eine lange Geschichte in der Prozess- und Verfahrenstechnik zurückblicken. Einfache Reglerstrukturen sind leichter zu realisieren und schon früh wurde die Frage gestellt, ob ein Mehrgrößensystem mittels mehrerer Eingrößenregler geregelt werden kann. Um hier eine zufrieden stellende Regelgüte erreichen zu können, muss entschieden werden, welche Ausgänge mit welchen Eingängen gekoppelt werden. Die erste Arbeit die diese Fragestellung beantwortete (Bristol, 1966), war motiviert von typischen Fragestellungen zur Regelung von Mehrgrößensystemen in der Verfahrenstechnik. Ein alternativer Ansatz zu solchen sogenannten *Interaktionsmaßen* wurde von Niederlinski (1971) vorgestellt. Interaktionsmaße analysieren die zugrunde liegende Struktur eines Systems und beschäftigen sich mit der Frage, welche Struktur ein dezentraler Regler haben sollte, um eine gute Regelgüte zu erreichen. Der eigentliche Reglerentwurf ist davon unabhängig. Die vorliegende Arbeit beschäftigt sich in Kapitel 3 mit dem Entwurf von dezentralen Reglern. Grundsätzlich besteht der dezentrale Reglerentwurf aus zwei Teilproblemen: Erstens muss die Struktur des dezentralen Reglers festgelegt werden und zweitens muss der Regler selbst bzw. dessen Dynamik entworfen werden. Im Unterschied zu den oben beschriebene Interaktionsmaßen und anderen Ansätzen zur dezentralen Regelung (Rotkowitz and Lall, 2006; Shah and Parrilo, 2008; Qi et al., 2004; Scherer, 2002) werden in dieser Arbeit Reglertopologie und –dynamik nicht als zwei getrennte Teilprobleme betrachtet sondern *gleichzeitig* optimiert. Die Frage von Reglertopologie und –dynamik wird also nicht in zwei aufeinander folgenden Schritten getrennt beantwortet, sondern innerhalb *eines* Optimierungsproblems. Hierbei kann zwischen dem Dezentralisierungsgrad des Reglers und der erreichbaren Regelgüte abgewägt werden. Das resultierende Optimierungsproblem ist kombinatorischer Natur und NP-hart. Daher widmet sich ein Großteil dieses Kapitels der Approximation der ursprünglichen Fragestellung durch konvexe Optimierungsprobleme. Insbesondere Methoden aus dem Bereich des Compressive Sensing werden hier adaptiert um dünnbesetzte Reglermatrizen mit hohem Dezentralisierungsgrad zu entwerfen.

Topologieentwurf für vernetzte Systeme ist im Vergleich zur dezentralen Regelung eine neuere Fragestellung in der Regelungstechnik. Bei den hier beschriebenen Problemen gibt es eine starke Verbindung zwischen dem dynamischen Verhalten des Systems und dem Graph, der die Topologie beschreibt (Mesbahi and Egerstedt, 2010). Daher ist es möglich, die Eigenschaften des geschlossenen Kreises durch eine geeignete Wahl des zu Grunde liegenden Graphen zu beeinflussen. Die Graphentheorie ist deshalb das geeignete mathematische Werkzeug um diese Art von Systemen zu analysieren und zu entwerfen. Im Besonderen können homogene Systeme durch ungewichtete Graphen beschrieben werden und viele analytische Resultate existieren, die dynamische Eigenschaften mit der Topologie des Graphen verknüpfen, z.B. hängt die Konvergenzgeschwindigkeit eines Konsenzproblems damit zusammen, wie stark zusammenhängend ein Graph ist (Fax and Murray, 2004). Leider existieren für gewichtete Graphen, die heterogene Systeme beschreiben, kaum analytische Ergebnisse. Hier ist man beim Entwurf der Graphen auf optimierungsbasierte Methoden angewiesen. Dasselbe gilt, wenn klassische regelungstheoretische Gütekriterien wie z.B. $\mathcal{H}_\infty$- oder $\mathcal{H}_2$-Regelgüte eines vernetzten Systems betrachtet werden. In Kapitel 4 werden optimierungsbasierte Methoden vorgestellt, die es erlauben für verschiedene Arten von vernetzten Systemen die Topologie zu entwerfen. Ergebnisse aus der Graphentheorie und der Systemtheorie werden verknüpft um mit Hilfe von semidefiniter Programmierung optimale Graphen zu entwerfen. Der Schwerpunkt liegt hierbei auf einem möglichst dünnbesetzten Graphen, d.h. die Graphen sollen Gütekriterien einhalten mit garantierter Robustheit gegenüber Modellunsicherheiten, und dabei möglichst wenig Kanten haben. Dünnbesetzte Graphen haben wenig Kanten die die einzelnen Knoten verbinden. Wenn der Graph die Topologie eines vernetzten dynamischen

Systems beschreibt, bedeutet dies, dass es nur wenig Verbindungen zwischen den einzelnen Agenten gibt und daher nur wenig Informationsaustausch notwenig ist. Die entstehenden Optimierungsprobleme sind wie im Fall der dezentralen Regler kombinatorischer Natur (eine Kante ist vorhanden oder nicht). Wie im Fall des dezentralen Reglerentwurfs, verwenden wir Methoden aus dem Bereich des Compressive Sensing um die kombinatorischen Optimierungsprobleme durch konvexe Optimierungsprobleme zu approximieren.

Damit lässt sich der Hauptbeitrag der Arbeit zusammenfassen: Methoden aus dem Bereich der Signalverarbeitung, insbesondere des Compressive Sensing (Candès et al., 2006b,a; Donoho, 2006) und der Rekonstruktion dünnbesetzter *(engl. sparse)* Signale, sollen für die Regelungstheorie nutzbar gemacht werden.

Ein Vektor oder eine Matrix heißen dünnbesetzt, wenn die meisten ihrer Elemente Null sind. Dünnbesetzte Vektoren und Matrizen können effizient gespeichert werden, indem nur Position und Wert der Einträge ungleich Null gespeichert werden. Die Auswertung eines dünnbesetzten Matrix-Vektor-Produkts kann ebenfalls effizient erfolgen, da die Nullen in der Berechnung des Produkts nicht berücksichtigt werden müssen. Die 0-Norm ist ein Maß dafür, wie dünnbesetzt ein Vektor oder eine Matrix ist. Sie gibt die Anzahl der Elemente an, die ungleich Null sind. Dünnbesetzte Vektoren durch eine Optimierung über die 0-Norm zu finden ist jedoch ein schwieriges kombinatorisches Optimierungsproblem. Eine vollständige Suche über alle Kombinationen ist in den meisten Fällen die einzige Möglichkeit, das globale Optimum zu finden (Candès et al., 2006b). Im Bereich des Compressive Sensing werden dünnbesetzte Signale rekonstruiert, indem die 0-Norm durch die konvexe 1-Norm ersetzt wird. Das überraschende an dieser Substitution ist, dass ihre Effektivität theoretisch begründet werden kann, d.h. unter bestimmten Umständen löst eine ℓ_1-Minimierung das ℓ_0-Problem mit hoher Wahrscheinlichkeit. Der Grund hierfür ist, dass die 1-Norm eine konvexe Approximation der 0-Norm darstellt. In Fazel (2002) wurde mit Hilfe von konjugierten Funktionen gezeigt, dass die 1-Norm die konvexe Hülle der 0-Norm ist, und damit deren beste konvexe Approximation (im Sinne der Lagrangedualität).

Wie weiter oben beschrieben, treten kombinatorische Optimierungsprobleme auch im Bereich der Regelungstheorie auf. Insbesondere der Topologieentwurf für dezentrale Regler oder Netzwerke zählt dazu: eine Kopplung zwischen zwei Agenten ist vorhanden oder nicht, eine Messwert muss an einen (möglicherweise weit entfernten) Regler übermittelt werden oder nicht. Im Topologieentwurf ist wichtig, dass Anforderungen an die Regelgüte mit möglichst wenig solcher Kopplungen eingehalten werden, da einfache Strukturen weniger anfällig für Störungen sind und kostengünstiger implementiert werden können. Möglichst wenig Kopplungen zwischen Teilsystemen oder Teilsystem und dezentralem Regler werden durch dünnbesetzte Reglermatrizen und dünnbesetzte Graphen beschrieben. Um diese Strukturen zu erreichen wird in der vorliegenden Arbeit die ℓ_1-Minimierung mit Ergebnissen aus der Regelungtheorie verknüpft. Dadurch ist es möglich, mittels eines konvexen Optimierungsproblems die Topologie eines Netzwerkes oder eines Reglers zu optimieren und gleichzeitig Anforderungen an die Regelgüte als Nebenbedingungen zu berücksichtigen.

Die beschriebene Beziehung zwischen ℓ_0-Norm und ℓ_1-Norm wird noch einmal aus theoretischer Sicht in Kapitel 5 betrachtet. Ausgehend vom Erfolg des Compressive Sensing in der Signalverarbeitung gehen wir der Frage nach, ob sich das Konzept von dünnbesetzten Vektoren auf die Regelungstheorie und damit dynamische Systeme übertragen lässt. Eine ℓ_0-Verstärkung für dynamische Systeme, vergleichbar mit der $\mathcal{H}_2$- oder $\mathcal{H}_\infty$-Verstärkung aus der robusten Regelung, wird eingeführt. Wir beschreiben die ℓ_0-Verstärkung als das kleinste Verhältnis der Elemente ungleich Null zwischen Systemeingang und -ausgang. Es wird gezeigt, das diese Systemnorm durch die Anzahl der Elemente ungleich Null in der *Impulsantwort* des Systems charakterisiert ist. Ausgehend davon wird ein ℓ_0-Regelungsproblem eingeführt, und der Zusammenhang zwischen diesem und dem ℓ_1-Regelungsproblem gezeigt. Beim ℓ_1-optimalen Regelungsproblem (Dahleh and Khammash, 1993; Dahleh and Diaz-Bobillo, 1995) kann beobachtet werden, dass die Impulsantwort des geschlossenen Kreises dünnbesetzt ist. Aus dem oben beschriebenen Zusammenhang zwischen ℓ_0-Minimierung und ℓ_1-Minimierung ist dieses Phänomen nachvollziehbar. Wir geben jedoch eine systemtheoretische Erklärung für die dünnbesetzten Impulsantworten von ℓ_1-optimalen Reglern indem wir zeigen, dass das ℓ_1-optimale Regelungsproblem die beste konvexe Relaxierung des ℓ_0-Problems (im Sinne der Lagrangedualität) ist. Obwohl die Beiträge dieses Kapitels in erster Linie theoretischer Natur sind, ergeben sich direkte praktische Konsequenzen. Z.B. brauchen Filter mit endlicher Impulsantwort *(engl. finite impulse filters (FIR))* mit wenigen Elementen ungleich Null wenig Speicherplatz und arithmetische Operationen können schneller durchgeführt werden. Der hier vorgestellte Zusammenhang zwischen ℓ_1-optimaler Regelung und Filterentwurf mit dünnbesetzter Impulsantwort ist ein erster Schritt in diese Richtung.

Gliederung dieser Arbeit

Im Folgenden wird die Gliederung dieser Arbeit vorgestellt und eine kurze Zusammenfassung der einzelnen Forschungsbeiträge in den jeweiligen Kapitel gegeben.

Kapitel 2 – Sparsity Promoting Optimization (Optimierung bezüglich dünnbesetzer Strukturen)

In diesem Kapitel geben wir einen Überblick über Ergebnisse aus dem Bereich Compressive Sensing und der Rekonstruktion dünnbesetzter Signale. Wir vergleichen verschiedene Maße, die beschreiben wie dünnbesetzt ein Vektor ist. Der Schwerpunkt des Vergleichs liegt dabei auf der Verwendung der Maße als Zielfunktion in einem Optimierungsproblem, das numerisch effizient gelöst werden kann. Unser Ziel ist es, die Ergebnisse aus dem Bereich Compressive Sensing in die Regelungstheorie zu überführen, um dezentrale Regler und dünnbesetzte Graphen zu entwerfen.

Kapitel 3 – Decentralized Control of Interconnected Systems (Dezentrale Regelung verkoppelter Systeme)

In diesem Kapitel wird ein neuer Ansatz zum Entwurf von dezentralen Reglern vorgestellt. Der Schwerpunkt liegt dabei auf dem *gleichzeitigen* Entwurf von Reglertopologie und –dynamik.

- Der Dezentralisierungsgrad eines Reglers wird mit Hilfe der Besetzungsstruktur des zentralen Reglers formuliert. Wir formulieren ein Optimierungsproblem, das die Topologie und die Dynamik des Reglers gleichzeitig hinsichtlich der gestellten Anforderungen an die Regelgüte optimiert. Es werden Lösungsansätze für Zustandsrückführung sowie statische und dynamische Ausgangsrückführung vorgestellt.

- Das neuartige Optimierungsproblem ist kombinatorischer Natur mit nichtkonvexen Nebenbedingungen. Mit Hilfe der gewichteten ℓ_1-Relaxierung und einer Systemerweiterung können wir eine konvexe Approximation herleiten. Die vorgestellten konvexen Optimierungsalgorithmen basieren auf der iterativen Lösung von linearen Matrixungleichungen und lösen die zuvor vorgestellte Problemstellung. Ein weiterer Schwerpunkt liegt dabei auf der Wahl der Anfangswerte der Algorithmen um deren Konvergenzwahrscheinlichkeit zu erhöhen.

- Der vorgestellte Ansatz zum Entwurf dezentraler Regelstrukturen wird auf ein Energiesystem angewendet mit dem Ziel Schwingungen zwischen den Teilsystemen zu dämpfen. Mittels der vorgestellten Algorithmen ist es möglich, die Signale zu identifizieren, die am meisten zur Schwingungsdämpfung beitragen. Die Regelgüte des zentralen Reglers kann erreicht werden, mit nur etwa 10% der ursprünglichen Messgrößen. Eine Robustheitsanalyse zeigt, dass die entworfenen dezentralen Regler auch bei Schwankungen im Netz die Schwingungen erfolgreich dämpfen können.

Teile dieses Kapitels basieren auf Schuler et al. (2012a, 2013a).

Kapitel 4 – Sparse Topology Design for Dynamical Networks (Strukturoptimierter Topologieentwurf für dynamische Netzwerke)

Dieses Kapitel beschäftigt sich mit dem Topologieentwurf vernetzter Systeme. Hierbei werden zwei Modellklassen beschrieben: Erstens, das Konsenzprotokoll als kanonisches Beispiel für vernetzte Systeme und zweitens Netzwerke bei denen den Teilsysteme nur relative Messungen *(engl. relative sensing networks (RSN))* zur Verfügung stehen.

- Bei Konsenznetzwerken betrachten wir den Fall, bei dem zu einem existierenden Baum eine festgelegte Anzahl von Kanten hinzugefügt werden soll, so dass die Verbesserung der Regelgüte maximal ist. Das dadurch entstehende gemischt-ganzzahlige Optimierungsproblem wird in ein numerisch attraktives konvexes Optimierungsproblem umgewandelt.

- Für Netzwerke, denen als Messungen nur relative Größen zur Verfügung stehen, stellen wir ein Optimierungsproblem für den Topologieentwurf vor. Ein Schwerpunkt liegt dabei darauf, dass der entworfene Graph nur möglichst wenig Kanten enthält aber die Ansprüche an die Regelgüte trotzdem erfüllt werden. Wir betrachten hierbei nominelle Netzwerke und Netzwerke die robust gegenüber Modellunsicherheiten sind. Auch hier stellen wir attraktive Optimierungsalgorithmen auf der Basis von konvexen Matrixungleichungen vor, mit denen Graphen entworfen werden können, die die Anforderungen an die Regelgüte erfüllen.

In diesem Kapitel gibt es kein eigenständiges Beispielkapitel. Anstelle dessen werden verschiedene Beispiele innerhalb der einzelnen Unterkapitel betrachtet. Dieses Kapitel basiert hauptsächlich auf (Zelazo et al., 2012, Kapitel IV und V), (Zelazo et al., 2013, Kapitel 5 und 6), Schuler et al. (2012b) und Schuler et al. (2013c).

Kapitel 5 – ℓ_0-Gain and ℓ_1-Optimal Control (ℓ_0-Verstärkung und ℓ_1-optimale Regelung)

In diesem Kapitel wird die Definition von dünnbesetzten Vektoren auf Signale und dynamische Systeme übertragen. Wir definieren dünnbesetzte dynamische Systeme und zeigen mit Hilfe der Lagrangedualität warum die Impulsantwort eines ℓ_1-optimal geregelten Systems dünnbesetzt ist.

- Wir führen die ℓ_0-Norm für Signale und eine ℓ_0-Verstärkung für dynamische Systeme als Erweiterung der 0-Norm für Vektoren ein. Mit Hilfe der ℓ_0-Verstärkung formulieren wir das ℓ_0-Regelproblem.

- Mit Hilfe der Lagrangedualität zeigen wir dass das ℓ_0-Regelproblem und das ℓ_1-Regelproblem dasselbe duale Optimierungsproblem haben. Mit Hilfe dessen ist es uns möglich zu erklären warum die Impulsantwort des geschlossenen Kreises von ℓ_1-optimalen Reglern hauptsächlich Nullen enthält.

Die Ergebnisse dieses Kapitels basieren auf Schuler et al. (2011a).

Kapitel 6 – Conclusion (Fazit)

In diesem Kapitel werden die in der Arbeit beschriebenen Ergebnisse zusammen gefasst und in einen größeren Zusammenhang gesetzt. Des Weiteren zeigen wir Fragestellungen für mögliche zukünftige Forschungsarbeiten auf.

Um die Arbeit eigenständig lesbar zu machen, gibt es verschiedenen Anhänge auf die an den entsprechenden Stellen verwiesen wird.

1. Introduction

1.1. Motivation

The analysis and control of interconnected systems is one of the big challenges of modern engineering science (see Murray, 2002). Applications range from from highly coupled chemical plants, heat exchangers, chemical reaction networks and power generation networks to communication networks like the internet, transport systems consisting of many cars, trains and airplanes or environmental surveillance by unmanned aircrafts. The number of these systems is steadily increasing. On the one hand there are individual subsystems that want to operate individually but are influenced or disturbed by the couplings to their neighboring systems and on the other hand groups of individual agents have to fulfill common goals. Independent of the application field the constituent parts of these systems are the individual dynamical subsystems, the couplings between these subsystems, their interconnection with the controller and the controller architecture.

With the increasing number of networked control system also the performance demands on these type of systems increase. In addition to stability of the closed loop system one often faces strong performance requirements such as fast convergence in consensus problems, optimality in terms of certain system norms or guaranteed robustness against uncertainties. To fulfill these demands, increasing communication between individual agents is required and couplings between different subsystems cannot be ignored any longer. However, communication is expensive since the individual agents and subsystems are often spatially distributed over a large area (e.g. continental Europe in the case of power systems control) and information exchange therefore limited. This naturally rises the question how to fulfill performance requirements while keeping communication costs small i.e., how to identify the significant quantities in a network that contribute most to a satisfying performance. This question has a long history in the control systems community. Simple controller architectures are of advantage and one is often interested in controlling a multivariable system by individual single input single output controllers instead of multivariable controllers. It is then of importance which output is coupled to which input by the SISO controllers. The first work on these preferable loop pairings (Bristol, 1966) was motivated by typical problems in multivariable process control, an alternative approach was for example presented by Niederlinski (1971). These interaction measures analyze the underlying structure of an interconnected system and answer therefore the question, how a good structure of the decentralized controller might look like. The design of the decentralized controller dynamics itself is independent thereof. In this thesis, we further contribute to the topic of design decentralized controllers, which emphasis of *joint* design of controller topology and dynamics.

Topology design for networked control system is a more recent problem discussed in the control systems community. Here, a deep connection between the dynamical behavior and the underlying properties of the graph describing the interconnections is observed (Mesbahi and Egerstedt, 2010). Therefore, by designing the underlying graph of a network, one can influence the performance of the network. Here the use of graph theory as a tool for analysis and synthesis is recognized as the correct mathematical abstraction to study these systems. Especially homogenous systems can be described by unweighted graphs and many analytic results exists, coupling e.g. connectedness of a graph to the convergence rate of a consensus system (Fax and Murray, 2004). However, when heterogenous systems are considered (described as node and edge-weighted graphs), only few analytic results exits. The same holds true if classical performance metrics such as $\mathcal{H}_\infty$ or $\mathcal{H}_2$-norm are considered. The aim of this thesis is to help filling this gap in the literature and provide optimization based methods to design network topologies for homogenous and heterogenous network topologies that fulfill desired performance objectives. Furthermore, our emphasis is on the design of sparse networks i.e., networks that fulfill these objectives with as few links as possible.

Decentralization is accompanied by a loss of performance compared to a centralized setup. A controller or network with an all-to-all coupling has the most knowledge of the system and can act accordingly. If we want to achieve the same performance with a sparse structure, it is in general not clear how a good structure should look like. Often a combinatorial search over all possible combinations of communication or measurement links is the only solution (see Candès et al., 2006a). However, in compressive sensing or sparse signal recovery sparse signals are reconstructed exactly from what appear to be highly incomplete sets of linear measurements by constrained ℓ_1 minimization. This is the main contribution of this thesis: By combining control theoretic insights withs results from compressive sensing (see e.g. Candès et al., 2006b,a; Donoho, 2006) we provide numerically efficient solutions to problems that are otherwise only solvable by exhaustive search. We provide new results to the topic of topology design of decentralized controllers and networks, where only few results exist so far. The topology is not fixed *a-priori* but a free parameter in the optimization problem. We further introduce a notion of sparsity for dynamical systems. Inspired by the idea of sparse signals in compressive sensing, we define an ℓ_0-gain for dynamic systems in the spirit of $\mathcal{H}_\infty$ and $\mathcal{H}_2$-gains in robust control. We show how this new system gain is connected to the design of sparse finite impulse response (FIR) filters when cast as an optimal control problem.

1.2. Contribution and Outline of this Thesis

The following overview presents the outline of this thesis and briefly summarizes its contributions.

Chapter 2 – Sparsity Promoting Optimization

In this chapter we review results from compressive sensing and sparse signal recovery. We compare different sparsity measures and discuss how they can be used to promote sparse controller and network topologies when used in optimization.

Chapter 3 – Decentralized Control of Interconnected Systems

In this chapter we introduce a new approach to design sparse decentralized controllers for interconnected systems. In particular we are interested in the *joint* design of controller topology and controller dynamics.

- We establish a new formulation of decentralization in terms of the sparsity of a centralized controller. This enables us to integrate the decentralization and performance constraints into one optimization problem to be solved simultaneously. We provide solutions for the state feedback, static and dynamic output feedback case.

- Numerically efficient optimization algorithms based on the iterative solution of linear matrix inequalities are formulated. Special emphasis is also put on how to choose initial values for the algorithms to make convergence to the global optimum more likely.

- The proposed design procedure for decentralized controller design is applied to design inter-area damping in power systems. We identify the most important feedback signals and are able to capture the performance of the optimal centralized controller with only 10% of the originally required feedback.

Parts of this chapter are based on Schuler et al. (2012a, 2013a).

Chapter 4 – Sparse Topology Design for Dynamical Networks

This chapter focuses on sparse topology design of networked dynamical systems. We consider two different scenarios: the consensus problem as a canonical problem in networked control systems and relative sensing networks as a special class of networks with more complex agent dynamics.

- We consider the consensus network. Given a tree topology, we provide a solution to the problem where to add an a-priori fixed number of edges, such that the performance improvement is maximized. For the resulting mixed-integer problem, we provide numerically attractive convex approximations.

- We design the topology of relative sensing networks subject to performance constraints with special emphasis on the sparsity of the topology. Additionally, we show how to design topologies that are robust in face of model uncertainties. We provide numerically attractive convex optimization algorithms that approximate the originally proposed non-convex optimization problem.

3

In this chapter we do not present a separate example section. Instead, we use several illustrating examples throughout the different sections. This chapter is mainly based on Zelazo et al. (2012, Section IV and V), Zelazo et al. (2013, Section 5 und 6), Schuler et al. (2012b) and Schuler et al. (2013c).

Chapter 5 – ℓ_0-Gain and ℓ_1-Optimal Control

In this chapter we take the notion of sparsity in compressive sensing and transfer it to the context of systems theory. We define sparsity for dynamical systems and give a mathematical explanation for the sparse impulse responses observed in ℓ_1-optimal control.

- We introduce an ℓ_0-signal norm and gain for dynamical systems as an extension of the ℓ_0-vector norm. Using the ℓ_0-gain we formulate the ℓ_0 control problem.

- By application of Lagrange duality we show that the ℓ_0-control problem and the ℓ_1-control problem share the same dual optimization problem and give a mathematical explanation of the sparse closed loop response observed in ℓ_1-optimal controllers.

Results of this chapter are based on Schuler et al. (2011a).

Chapter 6 – Conclusion

This final chapter provides some conclusive remarks, summarizing the thesis and hints to possible future directions of research.

Supplementary material is provided in several appendices, referenced at appropriate places, with the aim to make this thesis self-contained.

2. Sparsity Promoting Optimization

In this chapter we give a brief introduction into different sparsity measures and how sparsity promoting optimization can be used for structure and topology design in control. We introduce basic concepts and tools that are used in this thesis. This discussion is based on Hurley and Rickard (2009); Lin et al. (2013) and Boyd and Vandenberghe (2004, Chapter 6).

Generally speaking, in a sparse representation, a small number of coefficients or elements contains a large proportion of the energy or information. A vector is called sparse if its sparsity measure is small compared to the dimension of the vector, i.e. if most of its entries are zero or close to zero, depending on the measure. Sparse representation of signals is of fundamental importance in many fields such as blind source separation, parameter estimation or identification (Hurley and Rickard, 2009), compression, sampling and signal analysis (Candès et al., 2006b) or model reduction (Peeters and Westra, 2004). More recently, the concept of sparse controller and filter design was also considered in the control system community (Schuler et al., 2013a; Baran et al., 2010; Fardad et al., 2011; Lin et al., 2013). Sparse controllers are of interest in networked control systems and decentralized control, where the controllers are spatially distributed over a possibly large area. If the controller is sparse, only very few measurements have to be transmitted and computation and transmission costs can be reduced. Sparse filters offer the opportunity to omit arithmetic operations and the elimination and deactivation of circuits components. When a group of agents has to achieve a common goal or task a sparse network topology is of advantage since it reduces the cost of link couplings between the individual agents.

2.1. Sparsity Measures and Their Use in Optimization

The heuristic interpretation of sparsity given previously leads to several possible sparsity measures (see Hurley and Rickard (2009) for an overview). Which sparsity measure should be used depends heavily on the considered application. Typically used sparsity measures are listed in Table 2.1. All listed sparsity measures have in common that they are easy to use in analysis, i.e. it is not difficult to decide if a given vector or matrix is sparse. However, we are interested in sparsity promoting optimization. This means that we want to *optimize* over a sparsity measure while satisfying constraints from a given control or network design problem as e.g. performance constraints, stability of a control loop or connectedness of a network. In the following we will describe advantages and disadvantages of the different sparsity measures when used as penalty functions in optimization problems.

Table 2.1.: Commonly used sparsity measures.

Measure	Definition
ℓ_0	number of $x_i \vert x_i \neq 0$
$\ell_{(0,\varepsilon)}$	number of $x_i \vert \vert x_i \vert > \varepsilon$,
ℓ_1	$\sum_{i=1}^{n} \vert x_i \vert$
weighted ℓ_1	$\sum_{i=1}^{n} m_i \vert x_i \vert$, $m_i \geq 0$
ℓ_p	$(\sum_{i=1}^{n} \vert x_i \vert^p)^{\frac{1}{p}}$ for $0 < p < 1$
log-sum	$\sum_{i=1}^{n} \log(1 + \frac{\vert x_i \vert}{\varepsilon})$ $0 < \varepsilon \ll 1$

The 0-norm of a vector $x \in \mathbb{R}^n$ is defined as

$$\|x\|_0 = \{\text{number of } x_i \vert x_i \neq 0\},$$

and corresponds exactly to the number of non-zero entries in x. Despite not satisfying the definition of a norm (homogeneity is not fulfilled, i.e. $\|\alpha x\|_0 = \|x\|_0 \neq |\alpha| \cdot \|x\|_0$), it is often referred to one in the literature. This is the traditional sparsity measure in many mathematical settings: an element contains information ($x_i \neq 0$) or it contains no information ($x_i = 0$). The 0-norm is very sensitive to noise and numerical accuracy. Therefore, in noisy settings the 0-norm is often replaced by the $\ell_{(0,\varepsilon)}$-measure

$$\|x\|_{0,\varepsilon} = \{\text{number of } x_i \vert \vert x_i \vert > \varepsilon\},$$

where only elements with absolute value larger than a threshold ε are counted. Clearly, the solution then depends heavily on the choice of ε, which is different for each application. In an optimization problem, it might not be known in advance. Both, the ℓ_0 and the $\ell_{(0,\varepsilon)}$- measure, are difficult to use as penalty functions in optimization and result in non-convex optimization problems. In both cases, the gradient contains no information, since it is always zero when defined. Exhaustive combinatorial search is the only way to find the sparsest solution when the 0-norm or the $(0,\varepsilon)$-norm are considered (Candès et al., 2006a).

In the following, we want to approximate the 0-norm by functions that can be used more easily as penalty functions in optimization problems. For the two-dimensional case the level sets of the objective function and the feasible set of the constraints are shown in Figure 2.1. The penalty functions are depicted as a solid line and the feasible set (i.e. the constraints of the optimization problem) is depicted in gray surrounded by a dashed line. The solution of the optimization problem is sparse if the solid and the dashed line intersect on one of the axes, since then one of the elements in the solution vector is zero.

The 1-norm of a vector $x \in \mathbb{R}^n$

$$\|x\|_1 = \sum_{i=1}^{n} |x_i|$$

approximates the 0-norm and is easily calculated. When used as penalty function in an optimization problem, this leads to a *convex* objective function and does not increase the complexity of a convex optimization problem. Surprising about this replacement is the fact that it can be theoretically justified to be effective, i.e., under suitable assumptions, ℓ_1-minimization solves the ℓ_0-problem with high probability. It can be shown that the 1-norm is the best convex approximation of the 0-norm: Let a map f with $f : \mathbb{X} \to \mathbb{R}$ be given, where $\mathbb{X} \subseteq \mathbb{R}^n$. The convex envelope of f (on $\mathbb{X}$), denoted f_{env}, is defined as the point-wise largest convex function g such that $g(x) \leq f(x)$ for all $x \in \mathbb{X}$. Then the following lemma holds:

Lemma 2.1 (Fazel (2002)). *The convex envelope of the function* $f(x) = \|x\|_0$ *on* $\mathbb{X} = \{x \in \mathbb{R}^n | \|x\|_\infty \leq 1\}$ *is* $f_{\text{env}}(x) = \|x\|_1$.

In linear programming, minimization over the 1-norm achieves a sparse solution except for the pathological case when the constraints are parallel to the 1-ball (see Figure 2.1(a)). However, as can be seen in the figure this is not the case when the feasible set is convex. Here the intersection of the level set and the feasible set is in general not on one of the axes. To overcome this problem, *weighted* ℓ_1-minimization

$$\sum_{i=1}^{n} m_i |x_i|, \quad m_i \geq 0 \tag{2.1}$$

was introduced by Candès et al. (2008). In weighted ℓ_1-minimization, a weight on each element counteracts the magnitude of this element and enforces the sparsity of the achieved solution (see also Figure 2.1(b)). When the weight is chosen inversely proportional to the magnitude of this element,

$$\begin{cases} m_i = 1/|x_i|, & x_i \neq 0 \\ m_i = \infty, & x_i = 0, \end{cases} \tag{2.2}$$

the 0-norm and the 1-norm of the vector coincide. These two important properties, namely the convex formulation and the coincidence with the 0-norm for properly chosen weights make the weighted ℓ_1-optimization very attractive. As can be seen in Figure 2.1(d) these properties only hold for $p = 1$. Minimizations over ℓ_p-measures

$$\|x\|_p = \left(\sum_{i=1}^{n} |x_i|^p \right)^{\frac{1}{p}}$$

with $p > 1$ (e.g. $p = 2$ or $p = \infty$) are also convex problems but will in general not lead to sparse solutions. To analyze, if a vector or matrix is sparse, ℓ_p-norms for $0 < p < 1$ and log-sum measures (see Figure 2.1(c))

$$\sum_{i=1}^{n} \log(1 + \frac{|x_i|}{\varepsilon}), \quad 0 < \varepsilon \ll 1$$

are also often used. As can be seen, the level-sets of the p-norm for $0 < p < 1$ and the feasible set intersect on one of the axes, i.e. the achieved solution when used in optimization is sparse. The log-sum measure has also a star-like shape and

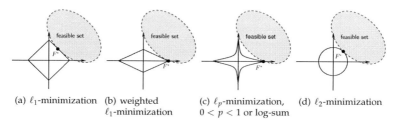

(a) ℓ_1-minimization (b) weighted ℓ_1-minimization (c) ℓ_p-minimization, $0 < p < 1$ or log-sum (d) ℓ_2-minimization

Figure 2.1.: The solution F^* of a constraint optimization problem is the intersection of the feasible set with the smallest sub-level set of the penalty function.

enforces sparsity outside some range. Both, the ℓ_p and the log-sum measures promote sparse solutions, however, they cannot be formulated as convex optimization problems when used as penalty functions.

Based on the previous discussion, we can conclude, that the weighted ℓ_1-minimization is the most favorable measure to promote sparsity in optimization problems. Unfortunately, the weighting scheme introduced in (2.2) cannot be implemented directly, since the weights depend on the solution of the optimization problem. In Candès et al. (2008) an iterative scheme is proposed, where the weights are the solution of the previous iteration (see Algorithm 2.1). In Fazel et al. (2003) it was

---

Algorithm 2.1 Re-weighted ℓ_1-minimization algorithm

1. Set $k = 1$. Choose a sufficiently small number $\nu > 0$ and initial weighs m_i^1.

2. Solve the minimization problem

$$\inf_{x^k} \sum_{i=1}^{n} w_i^k |x_i^k|$$

subject to constraints

3. Update the weights

$$w_i^{k+1} = \frac{1}{|x_i^k| + \nu},$$

where ν ensures that the inverse is always well defined.

4. Terminate on convergence. Otherwise set $k = k + 1$ and go to Step 2.

---

shown, that this so-called *re-weighted* ℓ_1-optimization belongs to the more general class of Majorization Minimization algorithm and is a special case of an iterative algorithm for minimizing the rank of a matrix subject to convex constraints. These algorithms converge to a local minimum but cannot be expected to always find the global minimum. Fast convergence of the re-weighted ℓ_1-minimization was shown for many numerical examples in Candès et al. (2008). To initialize the algorithm,

Candès et al. (2008) proposed to use the results of an ℓ_1-minimization, i.e. equal weights ($m_i^1 = 1$) on all elements. However, in the subsequent of this thesis, we will see that the initial weights can be used to steer the solution of the optimization problem into a preferred direction. If a certain sparsity pattern is desirable, this can be promoted by an appropriate choice of initial values m_i^1. Additionally, we show how control theoretic knowledge can be used when choosing m_i^1 to make convergence more likely.

2.2. Block Sparsity and Pattern Operator

The previous section discussed sparsity for vectors. In the following we will generalize this concept for matrices. Additionally, we will show, how to represent the sparsity pattern of a matrix and how we can deal with block matrices. The 0-norm of a matrix $A \in \mathbb{R}^{n \times m}$ is defined as

$$\|A\|_0 := \{\text{number of } a_{ij} | a_{ij} \neq 0\}.$$

Similar to the 0-norm of a vector, it gives the number of non-zero elements of a matrix and it holds that $\|A\|_0 = \|\text{vec}(A)\|_0$. However, more often, we are not only interested in the number of single zeros in a matrix, but rather in the number of *blocks of zeros* of predefined size in a matrix. Given a matrix $A \in \mathbb{R}^{n \times m}$, with

$$A = \begin{bmatrix} A_{11} & \cdots & A_{1M} \\ \vdots & \ddots & \vdots \\ A_{N1} & \cdots & A_{NM} \end{bmatrix}, \quad \text{and } A_{ij} \in \mathbb{R}^{n_i \times m_j} \text{ with } \sum_{i=1}^{N} n_i = n, \sum_{j=1}^{M} m_j = m$$

we define a *pattern operator* $\mathcal{Z}(A, p) \in \mathbb{R}^{N \times M}$

$$\mathcal{Z}(A, p) := \begin{bmatrix} \|A_{11}\|_p & \cdots & \|A_{1M}\|_p \\ \vdots & \ddots & \vdots \\ \|A_{N1}\|_p & \cdots & \|A_{NM}\|_p \end{bmatrix}, \quad p \in \{0, 1, 2, \infty\} \tag{2.3}$$

were each element z_{ij} of $\mathcal{Z}$ represents one block A_{ij} of A. Then, $z_{ij} = 0$ if and only if all elements of A_{ij} are zero. Using the pattern operator, we can represent the structure of a given matrix. The number of blocks that contain at least one non-zero number is given by $\|\mathcal{Z}(A, p)\|_0$.

The pattern operator is used in this thesis to represent the structure of a decentralized control scheme. Together with the re-weighted ℓ_1-minimization presented in the previous section we are able to design decentralized controllers and sparse networks as will be described in the next chapters.

3. Decentralized Control of Interconnected Systems

3.1. Overview

Interconnected systems are omnipresent in modern control engineering science (see Murray, 2002). The constituent parts of interconnected systems are the individual dynamical subsystems, the couplings between these subsystems, the interconnection with the controller and the controller architecture. In a decentralized framework, individual controllers are spatially distributed, and each controller has access to a different subset of local measurements. Decentralized control is often preferred to one centralized controller because less measurement wiring between sensors and controllers is necessary and less information needs to be transmitted (Šiljak, 1991). Today, distributed control of interconnected systems appears in a wide field of application areas such as distributed power generation, control of multi-agent systems, highly coupled chemical plants or chemical reaction networks. Due to higher demands on performance and efficiency, couplings between subsystems cannot be ignored anymore and have to be reflected in the controller structure.

In general, decentralized controller design consists of two steps: First, the structure of the decentralized controller has to be designed, and second, the controller itself has to be designed. In decentralized controller design, different levels of decentralization can be considered. Figure 3.1(a) shows exemplarily a network of three interconnected subsystems Σ_i, $i = 1, 2, 3$ with a centralized controller K that has access to and can influence all the subsystems. A centralized controller needs the measurement data of all subsystems. Therefore, analog or digital data connections are necessary. The (possibly) long distances between subsystems and the controller as well as the number of necessary connections are decisive for the realizability and the cost of the implementation of a controller. In a partially decentralized scheme (Figure 3.1(b)), some of the links between subsystems and controllers are omitted. This is often done to reduce implementation, communication and computational costs. In Figure 3.1(c), the network is controlled by three individual controllers K_{ii}, where each controller has only access to the measurements and control inputs of the associated subsystem. This is a completely decentralized control structure. The reduction of feedback control links is in general attended by a loss of performance compared to a centralized setup. But it is possible to achieve good control performance even when several feedback links are omitted in the control loop. This comes from the fact, that not all feedback links have the same influence on the performance of the closed loop.

While all this represents a general paradigm for the control of distributed systems and decentralized controller design, first approaches focusing on the design

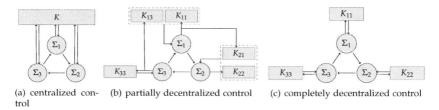

(a) centralized con-   (b) partially decentralized control   (c) completely decentralized control
trol

Figure 3.1.: Control schemes with different degree of decentralization.

of controller architecture and decentralized control were introduced in the process control community. For example the first work on preferable loop pairings (Bristol, 1966) was motivated by typical problems in multivariable process control, an alternative approach was for example presented by Niederlinski (1971). Generalizations and other interaction measures can among others be found in Gagnepain and Seborg (1982); Grosdidier and Morari (1986); Skogestad and Morari (1987); Johnston and Barton (1985); Luyben (1986) and Mijares et al. (1986). Recent work on loop pairing and interaction measures based on singular value decomposition and system Gramians include the work of He et al. (2006); Skogestad (2004) and Salgado and Conley (2004). Interaction measures analyze the underlying structure of an interconnected system and answer therefore the question, how a good structure of the decentralized controller might look like. They try to identify the loop pairings which are most important. The design of the decentralized controller itself is independent thereof. With this a completely decentralized controller as depicted in Figure 3.1(c) can be implemented. However, with such a controller structure, achievable performance generally decreases and sometimes not even stability of the interconnected system can be guaranteed.

With the advent of convex optimization and the efficient computation of $\mathcal{H}_\infty$ and $\mathcal{H}_2$-optimal controllers, decentralized controller design came again into focus and was considered in Rotkowitz and Lall (2006) and Shah and Parrilo (2008), as well as in Qi et al. (2004) and Scherer (2002). These approaches focus on the design of the $\mathcal{H}_\infty$ or $\mathcal{H}_2$-optimal controller dynamics and not on the design of the controller architecture. They have the following common features:

(i) the structure of the controller to be designed is fixed and restricted to special cases, and plant and controller have to share a common structure;

(ii) the structure of the controller has to be specified in advance, that is, they do not consider the problem of structure design but only the design of the controller dynamics.

Considering the problem of controller architecture and controller dynamics, a natural question one may raise is whether it is possible to design both controller structure and the controller dynamics *jointly*. This is especially of interest in the fast growing field of networked control systems with highly interacting subsystems, where it is often not clear, how the controller topology should look like to achieve good performance.

The above discussions motivate the main contribution of this chapter. We search for a tradeoff between the number of measurement links i.e., the degree of decentralization, and the achievable $\mathcal{H}_\infty$-performance of the closed loop system. In this sense, the controller topology is not specified in advance, but considered as an optimization variable. State feedback, static output feedback and dynamic output feedback are considered in the following. By means of a system augmentation approach, we first present a novel characterization of the $\mathcal{H}_\infty$-performance of the closed-loop system. Then, the non-convex structure optimization for topology design is relaxed by a convex weighted ℓ_1-minimization. The resulting problem can be tackled by finding a solution of iterative convex optimization problems. Moreover, we discuss how the initial values of the optimization can be chosen to improve the solvability of the proposed algorithm and propose algorithms to optimize initial values when necessary. Joint design of controller structure and controller dynamics subject to $\mathcal{H}_2$-performance constraint is also reported in Lin et al. (2013).

The proposed design procedure for decentralized controller design is applied to design inter-area damping in power systems. A tradeoff between necessary communication and achieved performance is computed. We show that only very few measurements have to be transmitted to either damp or disorder the inter-area modes and that the performance of the centralized optimal controller can be recaptured.

3.2. Interconnected Systems

We consider linear time-invariant systems of the form

$$
\Sigma_P : \quad
\begin{bmatrix} \dot{x} \\ z \\ y \end{bmatrix}
=
\begin{bmatrix} A & B_w & B_u \\ C_z & D_{zw} & D_{zu} \\ C_y & D_{yw} & D_{yu} \end{bmatrix}
\begin{bmatrix} x \\ w \\ u \end{bmatrix} ,
\tag{3.1}
$$

where $x \in \mathbb{R}^n$ is the system state, $w \in \mathbb{R}^{q_w}$ is the exogenous input (reference and disturbance), $u \in \mathbb{R}^{q_u}$ the control input, $z \in \mathbb{R}^{p_z}$ the performance output and $y \in \mathbb{R}^{p_y}$ the measurement output. Without loss of generality, we assume that $D_{yu} = 0$, since this can always be achieved by loop shifting (see Appendix A.2 for details). This system consists of N interconnected systems with system state $x_i \in \mathbb{R}^{n_i}$, $\sum_{i=1}^N n_i = n$ and $x = \left[x_1^T, \ldots, x_N^T \right]^T$. The state matrix A can then be decomposed according to the subsystem size, i.e.

$$
A = [A^{ij}] \text{ with } A^{ij} \in \mathbb{R}^{n_i \times n_j} \text{ and } i, j = 1 \ldots N.
$$

The other system matrices can also be decomposed into submatrices, e.g. $B_w = [B_w^{ij}]$ with $B_w^{ij} \in \mathbb{R}^{n_i \times q_{wj}}$, accordingly. In general, all system matrices can be full matrices without any structure. However, since we assume that system (3.1) consists of interconnected subsystems, in most cases we have $B_u = \text{diag}(B_u^{ii})$ and $C_y = \text{diag}(C_y^{ii})$, i.e. local actuators and local measurements, respectively. Additionally, the dynamic matrix A often possesses a (block-) sparse structure, i.e. we do not have an all-to-all coupling between the subsystems.

Assumptions on the model (3.1) are stabilizability of (A, B_u) and detectability of (A, C_y). In the following, we want to design different types of decentralized controllers for system (3.1), namely, state feedback controllers as well as static and dynamic output feedback controllers.

3.3. State Feedback

In the state feedback case, we assume that all states are measurable and available for feedback, i.e. $C_y = I$ and $D_{yw} = 0$ in equation (3.1). Performance is evaluated in terms of the $\mathcal{H}_\infty$-norm of the closed loop, where the performance output z is defined appropriately. In the state feedback case, an optimal controller can be designed and will be used as reference to evaluate the performance of the decentralized controller. The design procedure is outlined in the following.

Results of this chapter are based on Schuler et al. (2012a) and Schuler et al. (2013a), results for discrete-time systems can be found in Schuler et al. (2010a) and Schuler et al. (2010c).

3.3.1. Centralized Controllers

In a centralized framework the controller has access to all states. A state feedback controller is then given by

$$u = \hat{\mathcal{K}}x, \quad \hat{\mathcal{K}} \in \mathbb{R}^{q_u \times n}, \text{ with } \hat{\mathcal{K}} = [\hat{K}^{ij}] \text{ and } \hat{K}^{ij} \in \mathbb{R}^{q_{ui} \times n_j}. \tag{3.2}$$

In general, the matrix $\hat{\mathcal{K}}$ has no structure and the controller uses all possible measurement links between sensors and controller. We search for a controller that minimizes the influence of the exogenous input w on the performance output z in terms of the $\mathcal{H}_\infty$-norm of the closed loop. This centralized controller uses all possible degrees of freedom and can be designed via convex optimization (Gahinet and Apkarian, 1994). The closed loop is given by

$$\Sigma_{cl}(\hat{\mathcal{K}}): \quad \begin{bmatrix} \dot{x} \\ z \end{bmatrix} = \begin{bmatrix} A + B_u\hat{\mathcal{K}} & B_w \\ C_z + D_{zu}\hat{\mathcal{K}} & D_{zw} \end{bmatrix} \begin{bmatrix} x \\ w \end{bmatrix}. \tag{3.3}$$

We have the following assumptions on the centralized controller $\hat{\mathcal{K}}$ (3.2).

Assumption 3.1. *The centralized controller $\hat{\mathcal{K}}$ (3.2) achieves*

(i) *stability of the closed loop (3.3) and*

(ii) *optimality in terms of the $\mathcal{H}_\infty$-norm of the closed loop (3.3).*

3.3.2. Decentralized Controllers

Next, we describe a decentralized controller such that less measurement links between sensors and controllers are necessary. We especially consider that each controller does not have access to all subsystems but only knows the states of a few subsystems, i.e. we want to remove measurement links between subsystems and controllers. Therefore, we search for local controllers on the individual subsystems

$$u_i = \sum_{j=1}^{N} K^{ij} x_j, \quad K^{ij} \in \mathbb{R}^{q_{ui} \times n_j}$$

where $K^{ij} = 0$ for as many pairs (i, j) as possible. If $K^{ij} = 0$ no link from subsystem j to controller i is necessary. Combining all individual controllers into one state feedback controller similar to (3.2), leads to a decentralized controller

$$u = \mathcal{K}x, \quad \mathcal{K} \in \mathbb{R}^{q_u \times n}, \text{ and } \mathcal{K} = [K^{ij}]. \tag{3.4}$$

Similar to (3.3) the closed loop is given by

$$\Sigma_{cl}(\mathcal{K}) : \quad \begin{bmatrix} \dot{x} \\ z \end{bmatrix} = \begin{bmatrix} A + B_u \mathcal{K} & B_w \\ C_z + D_{zu} \mathcal{K} & D_{zw} \end{bmatrix} \begin{bmatrix} x \\ w \end{bmatrix}. \tag{3.5}$$

Note that $\mathcal{K}$ has a decentralized structure if it possesses many zero. In fact, decentralization of the controller is often modeled with a sparsity structure on the centralized controller, e.g. a completely decentralized control structure (see Figure 3.1(c)) would be represented by a (block) diagonal structure in the centralized controller $\hat{\mathcal{K}}$. The aim of decentralization is to reduce the number of non-zero elements of $\mathcal{K}$ such that only the important and effective elements remain. In general, we might be interested in different types of decentralization of the controller:

(i) Having as many zeros in $\mathcal{K}$, independent of their location in the controller matrix. This minimizes the communication that has to be transmitted over the network and

(ii) removing complete links between individual subsystems and local controllers, i.e. having as many K^{ij}-blocks equal to zero as possible. This is for example important if the individual subsystems are distributed over a wide area and the transmission of remote measurements is expensive.

Since $K^{ij} \in \mathbb{R}^{q_{ui} \times n_j}$, the second type of decentralization corresponds to zero matrices in the (i, j) block of the decentralized controller. To represent the structure of the decentralized controller, we use the pattern operator $\mathcal{Z}(\mathcal{K}, \infty)$ as defined in (2.3)

$$\mathcal{Z}(\mathcal{K}, \infty) := \begin{bmatrix} \|K^{11}\|_\infty & \cdots & \|K^{1N}\|_\infty \\ \vdots & \ddots & \vdots \\ \|K^{N1}\|_\infty & \cdots & \|K^{NN}\|_\infty \end{bmatrix}.$$

Remark 3.1. *The pattern operator above represents the block structure of the controller. To achieve this, not necessarily the ∞-norm has to be chosen. We are interested in a map that maps a block of zeros to a zero and any other block to a non-zero number. We are especially interested in a map that can be reformulated into a convex optimization problem and choose the ∞-norm for simplicity.*

With the pattern operator, we are able to evaluate the number of measurement data that is used by the controller $\mathcal{K}$. Each subcontroller K^{ij} is represented by a single element in the pattern operator and it holds that if $z^{ij} = 0$ no link from subsystem j to controller i is necessary. We will consider both types of decentralization in the following to optimize the structure of the controller.

3.3.3. Performance Degradation due to Decentralized Controller

As said before, the centralized controller can be computed via convex optimization and the global optimum can be found. Any decentralized controller will have the same or worse performance. In the present context, we allow a small performance degradation in terms of the $\mathcal{H}_\infty$-performance of the closed loop for the decentralized controller and want to use as few measurement links as possible to achieve this performance. Performance degradation due to the decentralization of the controller can now be investigated by the analysis of the error system $\Sigma_e(\hat{\mathcal{K}}, \mathcal{K}) = \Sigma_{cl}(\hat{\mathcal{K}}) - \Sigma_{cl}(\mathcal{K})$

$$\Sigma_e(\hat{\mathcal{K}}, \mathcal{K}) : \quad \begin{bmatrix} \dot{x}_e \\ e \end{bmatrix} = \begin{bmatrix} A_e & B_e \\ C_e & 0 \end{bmatrix} \begin{bmatrix} x_e \\ w \end{bmatrix}, \quad (3.6)$$

where

$$\left[\begin{array}{c|c} A_e & B_e \\ \hline C_e & D_e \end{array} \right] = \left[\begin{array}{cc|c} A + B_u\hat{\mathcal{K}} & 0 & B_w \\ 0 & A + B_u\mathcal{K} & B_w \\ \hline C_z + D_{zu}\hat{\mathcal{K}} & -(C_z + D_{zu}\mathcal{K}) & 0 \end{array} \right],$$

with $x_e = [\hat{x}\ x]^T$ and $e = \hat{z} - z$. Since the state variables of the error system $\Sigma_e(\hat{\mathcal{K}}, \mathcal{K})$ in (3.6) are composed of $\hat{x}$ and x, and the system $\Sigma_{cl}(\hat{\mathcal{K}})$ in (3.3) is asymptotically stable by Assumption 3.1, the asymptotic stability of $\Sigma_{cl}(\mathcal{K})$ in (3.5) is then equivalent to the asymptotic stability of $\Sigma_e(\hat{\mathcal{K}}, \mathcal{K})$. Thus, the requirement of the closed-loop system with the decentralized controller to be stable is achieved by guaranteeing the stability of the error system.

Note that the closed loop can also be written in the following way:

$$A_e = A_0 + FKR, \quad C_e = C_0 + GKR, \quad (3.7)$$

with

$$A_0 = \begin{bmatrix} A + B_u\hat{\mathcal{K}} & 0 \\ 0 & A \end{bmatrix}, \quad F = \begin{bmatrix} 0 \\ B_u \end{bmatrix}, \quad R = \begin{bmatrix} 0 & I \end{bmatrix}$$

$$C_0 = \begin{bmatrix} C + D_{zu}\hat{\mathcal{K}} & -C_z \end{bmatrix}, \quad G = -D_{zu}.$$

This representation of the error system resembles a static output feedback problem and will simplify the reformulation of the $\mathcal{H}_\infty$-performance $\Sigma_e(\hat{\mathcal{K}}, \mathcal{K})$ constraint in the next section.

We are now ready to formulate the decentralized control problem.

3.3.4. Decentralized Control Problem – State Feedback

In the following, we specify an acceptable performance degradation and minimize the 0-norm of the pattern operator $\mathcal{Z}(\mathcal{K}, \infty)$ in order to derive the optimally structured feedback. This can be formulated in the following decentralized control problem.

Problem 3.1 (Decentralized State Feedback Control Problem).
Given the linear system in (3.1) and the centralized controller in (3.2), determine a decentralized controller (3.4), such that

  (i) *the closed-loop system $\Sigma_{cl}(\mathcal{K})$ in (3.5) is asymptotically stable and*

 (ii) *the 0-norm of the pattern operator $\mathcal{Z}(\mathcal{K}, \infty)$ is minimized subject to a given maximal $\mathcal{H}_\infty$-performance degradation $\gamma > 0$, that is*

$$\min_{\mathcal{K} \ stabilizing} \left\| \mathcal{Z}(\mathcal{K}, \infty) \right\|_0 \tag{3.8a}$$

$$subject\ to\ \left\| \Sigma_{cl}(\hat{\mathcal{K}}) - \Sigma_{cl}(\mathcal{K}) \right\|_\infty < \gamma. \tag{3.8b}$$

Remark 3.2. *Note that instead of minimizing (3.8a) one can also use*

$$\min_{\mathcal{K} \ stabilizing} \left\| \mathcal{K} \right\|_0 \tag{3.9}$$

as an objective function. Then, the structure of the plant is not reflected in the achieved decentralization structure and the optimization just seeks to set as many elements as possible of the controller matrix to zero. This corresponds to the first type of decentralization described earlier.

The problem formulation implies that the performance degradation of the closed loop controlled with the decentralized controller does not exceed γ. Recall that the 0-norm of a vector is a measure of its sparsity. In this way, minimizing $\|\mathcal{Z}(\mathcal{K}, \infty)\|_0$ attempts to maximize the number of zero-elements of the pattern operator and therefore minimizes the number of measurement links. In the presented problem set-up, controller design and topology design are combined into one optimization problem and solved jointly.

As explained in Chapter 2, the minimization of the 0-norm is a non-convex problem and requires a combinatorial search (see Candès et al., 2006a). Furthermore, the $\mathcal{H}_\infty$-performance constraint is also non-convex and the well known convex reformulations cannot be applied to the presented set-up as will be seen in the next section. Therefore, Problem 3.1 is non-convex both in the objective function and in the constraint. In the following section, we show how this problem can be approximated by a numerically tractable convex optimization problem.

3.3.5. Convex Relaxation – State Feedback

This section focuses on the design of decentralized $\mathcal{H}_\infty$-controllers as defined in Problem 3.1. Our goal is to formulate a relaxed optimization problem, that can be solved by convex optimization techniques. We first use techniques as presented in Chapter 2 to approximate the numerically exhaustive combinatorial exact solution of the ℓ_0 objective function by a re-weighted ℓ_1-minimization. In a second step, we introduce a system augmentation approach to reformulate the characterization of the $\mathcal{H}_\infty$-performance of the closed loop system. This enables us to incorporate the decentralization constraint into the controller design.

Convex Relaxation of the ℓ_0-constraint

As described in Lemma 2.1, the convex envelope, and therefore the best convex approximation of the 0-norm, is the 1-norm. Furthermore, it was shown that the results achieved by ℓ_1-minimization can be further improved by *re-weighted* ℓ_1-minimization. In this direction, weights $m^{ij} \geq 0$ can be assigned to each variable K^{ij} and the non-convex objective function (3.8a) can be relaxed as

$$\min_{\mathcal{K} \text{ stabilizing}} \|\text{vec}(M \circ \mathcal{Z}(\mathcal{K}, \infty))\|_1, \tag{3.10a}$$

$$\text{subject to } \|\Sigma_{cl}(\hat{\mathcal{K}}) - \Sigma_{cl}(\mathcal{K})\|_\infty < \gamma. \tag{3.10b}$$

where $M = [m^{ij}]$ are non-negative weights and $\circ$ describes the Hadamard product between two matrices. For the described design problem, the weights are free parameters. They counteract the influence of the signal magnitude on the ℓ_1-penalty function. In the context of Problem 3.1, these weights can additionally be used to include system and control theoretic insight into the optimization problem to improve the result. For example, results from interaction measures (see e.g. Bristol, 1966; Grosdidier and Morari, 1986) can be used to choose appropriate initial weights. If other knowledge about the system is available, e.g. some measurement links are very unattractive since they are related to high implementation costs or just impossible to implement, those measurement links can also be penalized by a large initial weight in the re-weighted ℓ_1-minimization. Compared to Problem 3.1, the objective function is now convex. However, the non-convex $\mathcal{H}_\infty$-constraint remains and will be treated next.

Reformulation of the $\mathcal{H}_\infty$-performance Constraint

For the characterization of the $\mathcal{H}_\infty$-norm, the classical Bounded Real Lemma (B.1) formulation is used. Note that if matrix inequality (B.1) is used for controller synthesis, it is a bilinear matrix inequality due to the multiplication between the Lyapunov matrix P and the controller matrix $\mathcal{K}$ embedded in the closed loop matrices A_e and C_e and therefore it is non-convex. In existing convex approaches to $\mathcal{H}_\infty$-state feedback, the controller matrix does not appear explicitly in the synthesis conditions. This does not facilitate optimization with structural constraints as imposed on the controller in this thesis, where we are additionally interested in the minimization of $\|\mathcal{Z}(\mathcal{K}, \infty))\|_0$. For this reason, the approaches in Gahinet and Apkarian (1994), Sampei et al. (1990) or Iwasaki and Skelton (1994) are not directly applicable to the controller design problem discussed here. To overcome these difficulties, we introduce a system augmentation approach as initially developed in Shu and Lam (2009) for static output feedback and further adopted to positive filtering in Li et al. (2011). Such a formulation enables us to deal with the $\mathcal{H}_\infty$ decentralized control problem considered in this thesis.

We define an auxiliary variable

$$v = \mathcal{K}Rx_e \tag{3.11}$$

and consider $x_s^T = \begin{bmatrix} x_e^T & v^T \end{bmatrix}$ as an augmented variable, then we can write system (3.6) as a singular system

$$\Sigma_s : \quad \begin{bmatrix} E_s \dot{x}_s \\ e \end{bmatrix} = \begin{bmatrix} A_s & B_s \\ C_s & 0 \end{bmatrix} \begin{bmatrix} x_s \\ w \end{bmatrix}, \tag{3.12}$$

with

$$E_s = \begin{bmatrix} I & 0 \\ 0 & 0 \end{bmatrix}, \quad A_s = \begin{bmatrix} A_0 & F \\ \mathcal{K}R & -I \end{bmatrix}, \quad B_s = \begin{bmatrix} B_e \\ 0 \end{bmatrix}, \quad C_s = \begin{bmatrix} C_0 & G \end{bmatrix}.$$

In the following, we will present a theorem for stability and $\mathcal{H}_\infty$-performance of the singular system (3.12)

Theorem 3.1. *Given the decentralized controller $\mathcal{K}$, the following statements are equivalent*

(i) the error system in (3.6) is asymptotically stable and satisfies $\|\Sigma_e(\hat{\mathcal{K}}, \mathcal{K})\|_\infty < \gamma$,

(ii) there exist a symmetric matrix $P_1 > 0$ and diagonal matrix $P_2 > 0$ such that

$$\Omega := \begin{bmatrix} P_s^T A_s + A_s^T P_s & P_s^T B_s & C_s^T \\ B_s^T P_s & -\gamma I & 0 \\ C_s & 0 & -\gamma I \end{bmatrix} < 0, \tag{3.13}$$

with

$$P_s = \begin{bmatrix} P_1 & 0 \\ -\frac{1}{2} P_2 \mathcal{K}R & \frac{1}{2} P_2 \end{bmatrix}.$$

Proof. $(ii) \Rightarrow (i)$. Suppose there exist $P_1 > 0$ and diagonal $P_2 > 0$ such that (3.13) holds. Define a non-singular matrix

$$T = \begin{bmatrix} I & 0 & 0 & 0 \\ \mathcal{K}R & 0 & 0 & I \\ 0 & I & 0 & 0 \\ 0 & 0 & I & 0 \end{bmatrix}.$$

Pre- and post-multiplying (3.13) from left and right by T^T and T, respectively, we have

$$\bar{\Omega} := T^T \Omega T = \begin{bmatrix} P_1 A_e + A_e^T P_1 & P_1 B_e & C_e^T & P_1 F \\ B_e^T P_1 & -\gamma I & 0 & 0 \\ C_e & 0 & -\gamma I & G \\ F^T P_1 & 0 & G^T & -P_2 \end{bmatrix}.$$

The third leading principal submatrix of $\bar{\Omega}$ is identical to (B.1). Based on Lemma B.1 this indicates that the error system in (3.6) is asymptotically stable, and satisfies $\|\Sigma_e\|_\infty < \gamma$.

$(i) \Rightarrow (ii)$. If the error system in (3.6) is asymptotically stable, and satisfies $\|\Sigma_e(\hat{\mathcal{K}}, \mathcal{K})\|_\infty < \gamma$, then it follows from Lemma B.1 that there exists a matrix P_1, such that (B.1) holds.

Given any diagonal matrix $X > 0$, there must exist a scalar $\alpha > 0$ satisfying

$$- \alpha X - \begin{bmatrix} P_1 F \\ 0 \\ G \end{bmatrix}^T \begin{bmatrix} A_e^T P_1 + P_1 A_e & P_1 B_e & C_e^T \\ B_e^T P_1 & -\gamma I & 0 \\ C_e & 0 & -\gamma I \end{bmatrix}^{-1} \begin{bmatrix} P_1 F \\ 0 \\ G \end{bmatrix} < 0. \qquad (3.14)$$

By choosing $P_2 = \alpha X$ and applying the Schur complement (A.1) equivalence to (3.14), we have $\Omega = T^{-T} \bar{\Omega} T^{-1} < 0$, which completes the proof. $\qquad \square$

Remark 3.3. *The characterization of the $\mathcal{H}_\infty$-norm in Theorem 3.1 is necessary and sufficient and therefore equivalent to the formulation of Lemma B.1. The difference is the separation of the controller matrix $\mathcal{K}$ and the Lyapunov matrix P_1. Additionally, the controller matrix can be further parameterized by the diagonal matrix P_2. For the necessity part of the proof, the diagonal structure of P_2 is not needed (but there will always exist a P_2 with diagonal structure). But it will enable us later on to design the decentralized controller $\mathcal{K}$.*

The next theorem is derived from Theorem 3.1 and gives necessary and sufficient conditions for the solution of the decentralized control problem.

Theorem 3.2. *There exists a controller $\mathcal{K}$, if and only if there exist matrices $P_1 > 0$, diagonal $P_2 > 0$ and full block matrices L and U, such that*

$$\Pi(U) := \begin{bmatrix} \Delta & P_1 F + R^T L^T & P_1 B_e & C_0^T \\ F^T P_1 + LR & -P_2 & 0 & G^T \\ B_e^T P_1 & 0 & -\gamma I & 0 \\ C_0 & G & 0 & -\gamma I \end{bmatrix} < 0, \qquad (3.15)$$

with

$$\Delta = A_0^T P_1 + P_1 A_0 - U^T LR - R^T L^T U + U^T P_2 U.$$

In this case, the decentralized controller is given by

$$\mathcal{K} = P_2^{-1} L. \qquad (3.16)$$

Proof. Sufficiency: From (3.16), we have $L = P_2 \mathcal{K}$. Substituting this into (3.15), it is identical to (3.13) except for the first leading principle. By adding the nonnegative term $(U - \hat{\mathcal{K}} R)^T P_2 (U - \hat{\mathcal{K}} R)$ to the first principal, we observe that for any U it holds that

$$-R^T \mathcal{K}^T P_2 \mathcal{K} R \leq - R^T \mathcal{K}^T P_2 \mathcal{K} R + (U - \mathcal{K} R)^T P_2 (U - \mathcal{K} R)$$
$$= U^T P_2 U - U^T P_2 \mathcal{K} R - R^T \mathcal{K}^T P_2 U \qquad (3.17)$$

and the first leading principal of (3.15) is always greater than or equal than the first principal of (3.13). According to Theorem 3.1 this completes the sufficiency of the proof.

Necessity: Assume that the error system (3.6) is asymptotically stable with $\|\Sigma_e\|_\infty < \gamma$. Then, according to Theorem 3.1, there exist a matrix $P_1 > 0$ and a diagonal matrix $P_2 > 0$ such that (3.13) holds. By selecting $U = \mathcal{K}R$ we have

$$-R^T\mathcal{K}^TP_2\mathcal{K}R = -R^T\mathcal{K}^TP_2\mathcal{K}R + (U - \mathcal{K}T)^TP_2(U - \mathcal{K}R)$$
$$= U^TP_2U - U^TP_2\mathcal{K}R - R^T\mathcal{K}^TP_2U. \tag{3.18}$$

Substitute (3.18) into (3.13), and let $L = P_2\mathcal{K}$, it is equivalent to (3.15). This completes the proof. □

Remark 3.4. *Theorem 3.2 presents a necessary and sufficient condition for the existence of a decentralized controller. Although inequality (3.15) in Theorem 3.2 is still non-linear due to the multiplication of U, L, and U^TU one can see that the controller matrix $\mathcal{K}$ is only coupled with the diagonal matrix P_2, which is independent of the Lyapunov matrix P_1.*

With this new formulation for the $\mathcal{H}_\infty$-performance constraint we are now ready to formulate the relaxed decentralized control problem.

3.3.6. Relaxed Decentralized Controller Design – State Feedback

Based on the discussion of the previous section, we can relax Problem 3.1 with a convex objective function and the new characterization of the $\mathcal{H}_\infty$-performance. We therefore combine the weighted ℓ_1-minimization (3.10a) with (3.15).

Problem 3.2 (Relaxed State Feedback Decentralized Control Problem).
Given the linear system (3.1), find $P_1 > 0$, diagonal $P_2 > 0$, and full block matrices L and U satisfying

$$\inf \ \|\text{vec}(M \circ \mathcal{Z}(L, \infty))\|_1$$
subject to (3.15).

The decentralized controller is given by $\mathcal{K} = P_2^{-1}L$.

Due to the fact that P_2 is diagonal, the sparsity structure of $\mathcal{K}$ is preserved by specifying the sparsity of L. The advantage lies in the fact that the structural constraints added to $\mathcal{K}$ can now be imposed on L explicitly and it holds $\mathcal{Z}(\mathcal{K}, 0) = \mathcal{Z}(L, 0)$. The replacement of $\mathcal{K}$ by L in the 1-norm does change the objective function because $\|\text{vec}(M \circ \mathcal{Z}(\mathcal{K}, \infty))\|_1$ is not equivalent to $\|\text{vec}(M \circ \mathcal{Z}(L, \infty))\|_1$. However, this is not a substantial drawback because the relaxation from 0 to 1-norm anyway introduces some inaccuracy because entries with larger values are more penalized that entries with smaller values. Yet, this is compensated by the weighting matrix M.

As already stated in Remark 3.4, the matrix inequality (3.15) is generally not linear with respect to the parameters P_1, P_2, L and U. However, if we fix the matrix U, then it becomes an LMI in the other parameters. Thus, the remaining question is how to choose U properly. Inspired by Cao et al. (1998) it follows from (3.15) and (3.17) that for fixed P_1, P_2, L for all U, we have

$$\Pi(P_2^{-1}LR) \leq \Pi(U).$$

This implies that the minimal scalar $\varepsilon < 0$ satisfying $\Pi(U) < \varepsilon I$ is achieved for $U = P_2^{-1}LR$. This observation points to an iterative approach for solving (3.15). However, it is well known, that the convergence of an iterative algorithm depends heavily on the chosen initial conditions. A poor selection of a starting point often results in the iteration being trapped and no feasible solutions can be found, whereas a good selection may lead to a feasible solution. In the considered case, an initial value for U_1 has to be chosen.

We will use the fact that the solution to the centralized control problem is known and a natural choice is $U_1 = [0 \ \hat{\mathcal{K}}]$. This leads to a homotopy approach starting at the centralized solution and moving towards the decentralized controller. Based on the above discussion, we propose Algorithm 3.1 to solve Problem 3.2.

Algorithm 3.1 Relaxed decentralized controller design algorithm (State Feedback)

1. Set $k = 1$ and $U_1 = [0 \ \hat{\mathcal{K}}]$. Choose a sufficiently small number $\nu > 0$ and M_1.

2. For fixed U_k, solve the following convex optimization problem for the parameters $\Gamma = \{P_1 > 0, \text{ diagonal } P_2 > 0, L\}$:

$$\inf_{\Gamma} \|\text{vec}(M_k \circ \mathcal{Z}(L, \infty))\|_1$$

$$\text{subject to } \Pi(U_k) < 0.$$

3. Update U_{k+1} as $U_{k+1} = P_{2k}^{-1}L_k R$ and $m_{k+1}^{ij} = (|z_k^{ij}| + \nu)^{-1}$, where $\nu > 0$ ensures that the inverse is always well defined.
   Terminate on convergence or if $k = k_{\max}$ and go to Step 4 otherwise set $k = k + 1$ and go to Step 2.

4. Solve the following convex feasibility problem

$$\Pi(U_k) < 0$$

for the fixed controller structure obtained in Step 3. The controller is given as $\mathcal{K} = P_{2k}^{-1}L_k$.

Remark 3.5. *While the ℓ_1-relaxation leads to a semi-definite program that can be solved in polynomial time, there are still limitations to the solvable problem sizes. In practice, the algorithm performs well for problems with up to ~ 7000 variables (see e.g. the example in Chapter 3.6). Larger problems would require special solvers that take advantage of the structure of the individual problem.*

Remark 3.6. *In Step 4 of the algorithm, we compute a decentralized controller with the structure obtained in Step 3. This polishing step counteracts the fact, that the weighted*

ℓ_1-*minimization usually leads to small elements in the controller gain. This penalty for large values in $\mathcal{K}$ is removed in the last optimization.*

We adapted the systems augmentation approach as proposed in Shu and Lam (2009) and Li et al. (2011) to solve the decentralized controller design problem since it is applicable to a large system class. In contrast to other approaches from static output feedback such as e.g. Cao et al. (1998), Ghaoui and Oustry (1997), Leibfritz (2001), Mangasarian and Pang (1995) or Dinh et al. (2012) it can easily be adapted to discrete time systems and the number of additionally introduced variables is comparatively small. Good initial values for the iteration can be easily chosen in the state feedback case and in the case of static and dynamic output feedback algorithms for initial value optimization are provided (as will be seen in the following). As a most basic solution, a simple $(P, \mathcal{K})$-iteration would also be possible, but classical $(P, \mathcal{K})$-iteration has in general not very good convergence properties (see Simon et al. (2011) and referenced therein for details).

Even though convergence of the weighted ℓ_1-optimization to a local minimum was shown in Fazel et al. (2003), convergence of the proposed algorithm cannot be guaranteed, since additionally the constraint of the optimization problem is changed in each step of the iteration (update of U_k). However, we observed fast convergence for many numerical examples (see Schuler et al., 2012a; Schäfer, 2013) not presented in this thesis.

3.4. Static Output Feedback

In this section, we assume that instead of the whole state, only a part of the states is available for feedback, i.e. rank(C_y) < n and possibly $D_{yw} \neq 0$ in equation (3.1). This is often a more realistic scenario, since in almost all cases not all states are measurable. As a consequence state feedback is not possible and one has to rely on output feedback. In this section, we want to design *static* output feedback controllers. Static output feedback can be used to improve the performance of a locally implemented decentralized control structure by remote measurements without adding computational complexity. However, in contrast to optimal state feedback, there does not exists a convex solution to the optimal static output feedback problem and it is unknown if it is NP-hard (Blondel and Tsitsiklis, 2000). Therefore, the static output feedback problem without structure constraint is already very hard to solve. The subsequently presented approach for structured controller design does not increase the complexity of the problem while it achieves a decentralized structure in the controller.

Results of the presented setup for discrete-time systems can be found in Schuler et al. (2011b), details on the algorithmic properties and numerical issues can also be found in Schäfer (2013).

We want to design a *static output feedback* controller of the form

$$u_i = \sum_{j=1}^{N} K^{ij} y_j, \quad K_{ij} \in \mathbb{R}^{p_{yi} \times n_i}.$$

Combining these individual controllers to one controller leads to

$$u = \mathcal{K}y, \quad \mathcal{K} = [K^{ij}] \in \mathbb{R}^{p_y \times n}. \tag{3.19}$$

The closed loop is then given by

$$\Sigma_{cl}(\mathcal{K}) : \quad \begin{bmatrix} \dot{x} \\ z \end{bmatrix} = \begin{bmatrix} A + B_u \mathcal{K} C_y & B_w + B_u \mathcal{K} D_{yw} \\ C_z + D_{zu} \mathcal{K} C_y & D_{zw} + D_{zu} \mathcal{K} D_{yw} \end{bmatrix} \begin{bmatrix} x \\ w \end{bmatrix}. \tag{3.20}$$

Similar to the state feedback problem, $\mathcal{K}$ has a decentralized structure if it possesses many zeros. If we are not only interested in the sparsity of the controller matrix, but in the *block* sparsity of the controller matrix, we again use the following pattern operator

$$\mathcal{Z}(\mathcal{K}, \infty) := \begin{bmatrix} \|K^{11}\|_\infty & \cdots & \|K^{1N}\|_\infty \\ \vdots & \ddots & \vdots \\ \|K^{N1}\|_\infty & \cdots & \|K^{NN}\|_\infty \end{bmatrix}.$$

3.4.1. Decentralized Control Problem – Static Output Feedback

As stated above, we cannot compute a globally optimal centralized controller. Therefore, we do not consider the error system as in the case of decentralized state feedback but directly try to minimize the $\mathcal{H}_\infty$-performance of the closed loop with the decentralized controller. This can be stated as follows:

Problem 3.3 (Decentralized Static Output Feedback Control Problem). *Given the linear system (3.1), determine a structured controller (3.19), such that*

(i) *the closed-loop system* $\Sigma_{cl}(\mathcal{K})$ *in (3.20) is asymptotically stable and*

(ii) *the 0-norm of the pattern operator* $\mathcal{Z}(\mathcal{K}, \infty)$ *is minimized subject to a given* $\mathcal{H}_\infty$-*performance constraint* $\gamma > 0$, *that is*

$$\min_{\mathcal{K} \text{ stabilizing}} \|\mathcal{Z}(\mathcal{K}, \infty)\|_0 \tag{3.21a}$$

$$\text{subject to } \|\Sigma_{cl}(\mathcal{K})\|_\infty < \gamma. \tag{3.21b}$$

Again, similar to the state feedback problem, it is also possible to directly consider $\|\mathcal{K}\|_0$ instead of $\|\mathcal{Z}(\mathcal{K}, \infty)\|_0$.

In the previously formulated optimization problem, we observe the same difficulties as in the state feedback case. Namely, the combinatorial nature of the objective function as well as the non-convex performance constraint. We will use similar approaches as in the state feedback case to derive a relaxed convex optimization problem.

3.4.2. Convex Relaxation – Static Output Feedback

The convex relaxation of the objective function is equivalent to the relaxation presented in Section 3.3.5 and we can replace the non-convex 0-norm minimization by a weighted ℓ_1-minimization

$$\min_{\mathcal{K} \text{ stabilizing}} \|\text{vec}(M \circ \mathcal{Z}(\mathcal{K}, \infty))\|_1, \tag{3.22a}$$

$$\text{subject to } \|\Sigma_{cl}(\mathcal{K})\|_\infty < \gamma. \tag{3.22b}$$

The constraint of the optimization problem is now a classical static output feedback problem and we can directly apply results from the literature. Again, we use the system augmentation approach as presented in Shu and Lam (2009) and Li et al. (2011), in this context without any adaptations. Define the auxiliary variable

$$v = \mathcal{K}C_y x + \mathcal{K}D_{yw}w \tag{3.23}$$

and consider $x_s^T = \begin{bmatrix} x^T & v^T \end{bmatrix}$ as an augmented variable, then we can write system (3.20) as a singular system similar to (3.12)

$$\Sigma_s : \quad \begin{bmatrix} E_s \dot{x}_s \\ z \end{bmatrix} = \begin{bmatrix} A_s & B_s \\ C_s & D_s \end{bmatrix} \begin{bmatrix} x \\ w \end{bmatrix}, \tag{3.24}$$

with

$$E_s = \begin{bmatrix} I & 0 \\ 0 & 0 \end{bmatrix}, \quad A_s = \begin{bmatrix} A & B_u \\ \mathcal{K}C_y & -I \end{bmatrix}, \quad B_s = \begin{bmatrix} B_w \\ \mathcal{K}D_{yw} \end{bmatrix}, \quad C_s = \begin{bmatrix} C_z & D_{zu} \end{bmatrix}, \quad D_s = D_{zw}.$$

The advantage of this description lies in the fact, that now the controller matrix $\mathcal{K}$ is exposed from the middle of two system matrices (namely B_u and C_y or D_{zu} and C_y, respectively) such that $\mathcal{K}$ is only attached to one further matrix.

We will now cite two theorems concerning the $\mathcal{H}_\infty$-performance of system (3.24):

Theorem 3.3 (Li et al. (2011)). *Given the controller $\mathcal{K}$ as in (3.19), the following statements are equivalent:*

(i) *The closed loop system Σ_{cl} in (3.20) is asymptotically stable and satisfies $\|\Sigma_{cl}\|_\infty < \gamma$,*

(ii) *there exist matrices $P_1 > 0$ and diagonal $P_2 > 0$ such that*

$$\begin{bmatrix} P_s A_s + A_s^T P_s & P_s^T(I+N)B_s & C_s^T \\ B_s^T(I+N)P_s & -B_s^T J^T(P_s P_s^T)J B_s - \gamma I & D_s T \\ C_s & D_s & -\gamma I \end{bmatrix} < 0, \tag{3.25}$$

with

$$P_s = \begin{bmatrix} P_1 & 0 \\ -\frac{1}{2}P_2\mathcal{K}C_y & \frac{1}{2}P_2 \end{bmatrix}, \quad N = \begin{bmatrix} 0 & 0 \\ 0 & I \end{bmatrix}.$$

Theorem 3.4 (Li et al. (2011)). *The decentralized control problem is solvable, if and only if there exits a matrix $P_1 > 0$, a diagonal matrix $P_2 > 0$ and a matrices U, V and L, such that*

$$
\Pi(U,V) := \begin{bmatrix} \Pi_{11} & P_1 B_u + C_y^T L^T & \Pi_{13} & C_z^T \\ B_u^T P_1 + LC_y & -P_2 & LD_{yw} & D_{zu}^T \\ \Pi_{13}^T & D_{yw}^T L^T & \Pi_{33} & D_{zw}^T \\ C_z & D_{zu} & D_{zw} & -\gamma I \end{bmatrix} < 0 \qquad (3.26)
$$

with

$$
\begin{aligned}
\Pi_{11} &= A^T P_1 + P_1 A - U^T LC_y - C_y^T L^T U + U^T P_2 U \\
\Pi_{13} &= P_1 B_w - C_y^T L^T V + U^T P_2 V - U^T LD_{yw} \\
\Pi_{33} &= -V^T LD_{yw} - D_{yw}^T L^T V + V^T P_2 V - \gamma I.
\end{aligned}
$$

In this case, the decentralized controller is given by

$$
\mathcal{K} = P_2^{-1} L.
$$

Inequality (3.26) offers now the possibility to include the topology of $\mathcal{K}$ indirectly into the synthesis process in the same fashion as for the state feedback case. This is possible due to the matrix L that can be used during controller synthesis, because it is independent of P_1 and P_2. Again, by choosing P_2 diagonal, the topology of L is preserved in $\mathcal{K}$. Moreover, it holds that $\mathcal{Z}(\mathcal{K},0) = \mathcal{Z}(L,0)$ and the sparsity pattern of $\mathcal{K}$ and L is the same. Therefore, the decentralization constraint can be reformulated as

$$
\|\mathrm{vec}(M \circ \mathcal{Z}(L,\infty))\|_1, \qquad (3.27)
$$

and the relaxed decentralization problem can be formulated as follows.

Problem 3.4 (Relaxed Static Output Feedback Decentralized Control Problem). *Given the linear system (3.1), find $P_1 > 0$, diagonal $P_2 > 0$, and full block matrices L, U and V satisfying*

$$
\inf \ \|\mathrm{vec}(M \circ \mathcal{Z}(L,\infty))\|_1
$$
$$
\textit{subject to (3.26).}
$$

The decentralized controller is given by $\mathcal{K} = P_2^{-1} L$.

Inequality (3.26) is still nonlinear due to the multiplication of P_1, P_2, U, V and L. However, if we fix U and V it becomes an LMI. Thus, the remaining question is how to choose U and V properly. As shown in Li et al. (2011) and Cao et al. (1998), for any fixed P_1, P_2, L, and U and V, we have

$$
\Pi\left(P_2^{-1} LC_y, P_2^{-1} LD_{yw} \right) \leq \Pi\left(U, V \right).
$$

This implies that the minimal scalar $\varepsilon < 0$ satisfying

$$\Pi(U, V) < \varepsilon H \quad \text{with} \quad H = \begin{bmatrix} I & 0 & 0 & 0 \\ 0 & 0 & 0 & 0 \\ 0 & 0 & I & 0 \\ 0 & 0 & 0 & 0 \end{bmatrix}$$

is achieved for $U = P_2^{-1}LC_y$ and $V = P_2^{-1}LD_{yw}$. Inspired by Cao et al. (1998) this observation points to an iterative approach for solving (3.26) as presented in Algorithm 3.2.

Algorithm 3.2 Relaxed decentralized controller design algorithm (Static output feedback)

---

1. Set $k = 1$ and choose initial values for U_1 and V_1. Choose a sufficiently small number $\nu > 0$, M_1 and set $\alpha > 0$.

2. For fixed U_k and V_k, solve the following convex optimization problem for the parameters $\Gamma = \{P_1 > 0, \text{ diagonal } P_2 > 0, L, \varepsilon_k\}$:

$$\inf_{\Gamma} \|\text{vec}(M_k \circ \mathcal{Z}(L, \infty))\|_1 + \alpha\varepsilon_k$$

$$\text{subject to } \Pi(U_k, V_k) < \varepsilon_k I.$$

3. If $\varepsilon_k < 0$, update U_{k+1} as $U_{k+1} = P_{2k}^{-1}L_kC_y$, $V_{k+1} = P_{2k}^{-1}L_kD_{yw}$ and $m_{k+1}^{ij} = (|z_k^{ij}| + \nu)^{-1}$, where $\nu > 0$ ensures that the inverse is always well defined. Terminate on convergence or if $k = k_{\max}$ and go to Step 4. Otherwise set $k = k + 1$ and go to Step 2.
   If $\varepsilon_k > 0$, set U_k as $U_k = P_{2k}^{-1}L_kC_y$ and V_k as $V_k = P_{2k}^{-1}L_kD_{yw}$ go to Step 2.

4. Solve the following convex feasibility problem

$$\Pi(U_k, V_k) < 0$$

for the fixed controller structure obtained in Step 3. The controller is given as $\mathcal{K} = P_{2k}^{-1}L_k^*$.

Remark 3.7. *In contrast to Algorithm 3.1, Algorithm 3.2 does not start within the feasible set of the LMI constraint. Therefore, we cannot solve $\Pi(U_k, V_k) < 0$ directly, but have to converge to the feasible set from the outside. This is done by considering a convex sum $\|\text{vec}(M_k \circ \mathcal{Z}(L))\|_1 + \alpha\varepsilon_k$ as an objective function, where we try to minimize both, the structure constraint and the LMI constraint at the same time. The weighting matrix M_k is only updated, if a feasible controller (i.e. $\varepsilon_k < 0$) is found. Note that the introduction of the penalty parameter α in Step 2 is due to the fact that we aim to minimize both $\|\text{vec}(M_k \circ \mathcal{Z}(L, \infty))\|_1$ and ε_k. For this algorithm no information on the convergence of the algorithm can be concluded. However, we received good convergence results in various examples not presented in this thesis (Schäfer, 2013).*

In Step 1, one can utilize existing approaches to compute U_1 and V_1. In fact, according to Lemma B.1, it is not difficult to verify that finding U_1 and V_1 is equivalent to solving

$$\begin{bmatrix} AQ + B_u W_1 + QA^T + W_1^T B_u^T & B_w + B_u V_1 & QC_z^T + W_1^T D_{zu}^T \\ B_w^T + V_1^T B_u^T & -\gamma I & D_{zw}^T + V_1^T D_{zu}^T \\ C_z Q + D_{zu} W_1 & D_{zw} + D_{zu} V_1 & -\gamma I \end{bmatrix} < 0 \qquad (3.28)$$

with $Q > 0$. U_1 can then be calculated as $U_1 = WQ^{-1}$. These initial values for U_1 and V_1 can then be used to initialize Algorithm 3.2. However, this is in general a poor choice for initial values and it is well known that the convergence of an iterative algorithm heavily depends on a good choice of initial values. The poor choice results from the fact that inequality (3.28) can be seen as a state feedback $\mathcal{H}_\infty$ controller synthesis problem which uses all system states for controller design. Whereas the original performance constraint should be solved by an output feedback controller, which has only access to a subset of the system states. We adapt now the idea from Li et al. (2011) to generate initial values for U_1 and V_1 such that the state feedback controller mainly uses system states which are also available for output feedback.

3.4.3. Initial Value Optimization – Static Output Feedback

First, consider the initial closed loop system

$$\Sigma_{cl,init}(U_1, V_1) : \quad \begin{bmatrix} \dot{x} \\ z \end{bmatrix} = \begin{bmatrix} A_{cl,init} & B_{cl,init} \\ C_{cl,init} & D_{cl,init} \end{bmatrix} \begin{bmatrix} x \\ w \end{bmatrix}, \qquad (3.29)$$

with

$$A_{cl,init} = A + B_u U_1$$
$$B_{cl,init} = B_w + B_u V_1$$
$$C_{cl,init} = C_u + D_{zu} U_1$$
$$D_{cl,init} = D_{zw} + D_{zu} V_1$$

by substituting $U = \mathcal{K} C_y$ and $V = \mathcal{K} D_{yw}$ in Σ_{cl} (3.20). Defining

$$\Phi = \begin{bmatrix} C_y & 0 & D_{yw} \end{bmatrix}$$
$$\Psi = \begin{bmatrix} U & 0 & V \end{bmatrix}$$

we can use the following theorem:

Theorem 3.5 (Li et al. (2011)). *Given Φ and Ψ as defined above, for a sufficiently small scalar $\varepsilon > 0$, the following statements are equivalent:*

(i) *There exists a controller $\mathcal{K}$ such that Σ_{cl} in (3.20) is asymptotically stable with $\|\Sigma_{cl}\|_\infty < \gamma$ and $\|\Psi - \mathcal{K}^* \Phi\|_2 < \varepsilon$,*

(ii) *$\Sigma_{cl,init}$ is asymptotically stable with $\|\Sigma_{cl,init}\|_\infty < \gamma$ and $\|\Psi \Phi^\perp\|_2 < \varepsilon$.*

The norm $\|\Psi\Phi^\perp\|_2$ defines how much of the space of Ψ coincidence with the null space of Φ. As Ψ represents the space of the system states used for control in state feedback and $\Phi^\perp$ the space of the system states that are not available for control due to output feedback, the coincidence of these two spaces should be as small as possible. Then it is more likely that the performance constraint (3.26) has a feasible solution and a stabilizing output feedback controller can be found. The norm condition $\|\Psi\Phi^\perp\|_2 < \varepsilon$ can be reformulated into the following LMI condition

$$\begin{bmatrix} -\varepsilon I & (\Psi\Phi^\perp)^T \\ \Psi\Phi^\perp & -I \end{bmatrix} < 0. \tag{3.30}$$

In addition, the proof of Theorem 3.5 in Li et al. (2011) delivers that

$$(\mathcal{A} + \mathcal{B}\Psi\mathcal{I})^T\mathcal{P} + \mathcal{P}(\mathcal{A} + \mathcal{B}\Psi\mathcal{I}) < 0 \tag{3.31}$$

with $P > 0$, $\gamma > 0$ and

$$\mathcal{A} = \begin{bmatrix} A & B_w & 0 \\ 0 & \frac{\gamma}{2}I & 0 \\ C_z & D_{zw} & \frac{\gamma}{2}I \end{bmatrix} \quad \mathcal{B} = \begin{bmatrix} B_u \\ 0 \\ D_{zu} \end{bmatrix} \quad \mathcal{I} = \begin{bmatrix} I & 0 & 0 \\ 0 & 0 & I \\ 0 & I & 0 \end{bmatrix} \quad \mathcal{P} = \begin{bmatrix} P & 0 & 0 \\ 0 & I & 0 \\ 0 & 0 & I \end{bmatrix}$$

is equivalent to inequality (3.28) by substituting $Q = P^{-1}$.

Due to the bilinearity it is not possible to minimize ε with respect to P and $\Psi(U, V)$. Therefore, the minimization is performed in two steps as an iterative LMI approach. Additionally to the initial value optimization as proposed in Li et al. (2011), γ is also included into the optimization problem. This is summarized in Algorithm 3.3.

Remark 3.8. *Note that in Algorithm 3.3, ε_k is monotonically decreasing with respect to k with a lower bound of zero (or β). In step 3, the convex sum $\gamma + \alpha\varepsilon$ is minimized subject to (3.31), (3.30) and $\varepsilon \geq \beta$. By choosing the scalar α sufficiently large it is possible to primaryly minimize ε. If ε has reached its lower bound, γ is further minimized with each iteration.*

Algorithm 3.3 and Algorithm 3.2 together provide a solution to the decentralized static output feedback problem stated in Problem 3.3. However, static output feedback often achieves not satisfactory performance of the closed loop. This is due to the fact, that the controller is only a static gain without dynamics. The next section focuses therefore on dynamic output feedback.

3.5. Dynamic Output Feedback

A natural extension to static output feedback is dynamic output feedback, where the controller is not only a gain but a dynamical system itself. Dynamic controllers have additional degrees of freedom and can often achieve a superior performance than static controllers. However, they are also computationally more demanding and need more memory when implemented. When no structural constraints are considered, there exist convex solutions to the dynamic output feedback problem

---

Algorithm 3.3 Intital value optimization

1. Set $k = 0$, $\alpha > 0$, $\beta > 0$, $\delta > 0$. Solve the convex optimization problem with the parameters $\Omega = \{Q > 0, \gamma_k > 0, W, V_k\}$

$$\inf_{\Omega} \gamma_k$$

subject to (3.28)

   to obtain $U_k = WQ^{-1}$ and V_k.

2. Set $k = k + 1$. For fixed U_k and V_k solve the convex optimization problem

$$\inf_{P_k, \gamma} \gamma$$

subject to (3.31)

   to obtain P_k.

3. For fixed P_k solve the convex optimization problem with the parameters $\Gamma = \{U_{k+1}, V_{k+1}, \gamma_k, \varepsilon_k\}$

$$\inf_{\Gamma} \gamma_k + \alpha \varepsilon_k$$

subject to (3.30)
(3.31)
$$\varepsilon_k \geq \beta$$

   to obtain U_{k+1}, V_{k+1} and γ_k

4. If $|\gamma_{k-1} - \gamma_k| < \delta$, where δ is a prescribed tolerance the initial values for U and V are obtained. Continue with Algorithm 3.2. Otherwise go to Step 2.

(see e.g. Gahinet and Apkarian, 1994). When structural constraints are added, convex solutions exist only for very special cases (see e.g Rotkowitz and Lall, 2006; Shah and Parrilo, 2008; Scherer, 2002). In the following, we give a solution to the decentralized dynamic output feedback problem that is not restricted to certain system classes. However, we can solve the problem only iteratively. Results of the presented setup for discrete-time systems can be found in Schuler et al. (2011c).

In a fully decentralized setup (see Figure 3.1(c)), for each subsystem $i = 1, \ldots, N$, a single dynamic output-feedback controller is assumed in the form of

$$\begin{aligned}
\dot{x}_{K,i} &= A_K^{ii} x_{K,i} + B_K^{ii} y_i \\
u_i &= C_K^{ii} x_{K,i} + D_K^{ii} y_i,
\end{aligned}$$

where $x_{K,i} \in \mathbb{R}^{n_{K,i}}$ is the state of the ith subcontroller and each subcontroller receives only measurements from its own subsystem. To improve the performance of these local dynamic controllers, we again search for a partially decentralized controller with a minimal number of additional measurement links between subsystems and controllers while keeping the $\mathcal{H}_\infty$-performance below a certain level γ, that is, we want to add measurements of other subsystems to the individual controllers. Therefore, we search for N controllers of the form

$$
\begin{aligned}
\dot{x}_{K,i} &= A_K^{ii} x_{K,i} + B_K^{ii} y_i \\
u_i &= C_K^{ii} x_{K,i} + \sum_{j=1}^N D_K^{ij} y_j,
\end{aligned}
\tag{3.32}
$$

with $x_{K,i}$ as introduced before and $D_K^{ij} = 0$ for as many pairs (i,j) as possible (see Figure 3.1(b)).

In this setting, each subsystem has a controller which may not only use the output of its own associated subsystem, but also selected outputs of other subsystems entering the feedthrough matrix D_K^{ij}. Note that $n_{K,i}$ can be chosen as $n_{K,i} \leq n$, which means that the ith subcontroller (3.32) can be of reduced order. If $n_{K,i} = 0$, then the ith subcontroller will degenerate into the static output feedback case, that is, $u_i(k) = \sum_{j=1}^N D_K^{ij} y_j(k)$. Therefore, the presented dynamic output feedback controller can also be seen as an extension to the static output feedback case presented in the previous section.

The controller with a decentralized structure is then given by

$$
\Sigma_K : \quad
\begin{bmatrix} \dot{x}_K \\ u \end{bmatrix}
=
\begin{bmatrix} A_K & B_K \\ C_K & D_K \end{bmatrix}
\begin{bmatrix} x_K \\ y \end{bmatrix},
\tag{3.33}
$$

where $x_K(k) = [x_{K,1}^T(k), \ldots, x_{K,N}^T(k)]^T$, A_K, B_K and C_K have a block diagonal structure and D_K defines the decentralized structure of the controller.

Let $\zeta = [x^T, x_K^T]^T$, then the closed loop with decentralized controller can be written as

$$
\Sigma_{cl}(\Sigma_K) : \quad
\begin{bmatrix} \dot{\zeta} \\ z \end{bmatrix}
=
\begin{bmatrix} A_{cl} & B_{cl} \\ C_{cl} & D_{cl} \end{bmatrix}
\begin{bmatrix} \zeta \\ w \end{bmatrix}
\tag{3.34}
$$

where

$$
\begin{aligned}
A_{cl} &= \begin{bmatrix} A + B_u D_K C_y & B_u C_K \\ B_K C_y & A_K \end{bmatrix}, &
B_{cl} &= \begin{bmatrix} B_w + B_u D_K D_{yw} \\ B_K D_{yw} \end{bmatrix}, \\
C_{cl} &= \begin{bmatrix} C_z + D_{zu} D_K C_y & D_{zu} C_K \end{bmatrix}, &
D_{cl} &= D_{zw} + D_{zu} D_k D_{yw}.
\end{aligned}
$$

Note, that the closed loop can be rewritten such that the controller appears as one parameter matrix $\mathcal{K}$:

$$
\begin{aligned}
A_{cl} &= A_0 + J\mathcal{K}R, & B_{cl} &= B_0 + J\mathcal{K}T, \\
C_{cl} &= C_0 + F\mathcal{K}R, & D_{cl} &= D_0 + F\mathcal{K}T,
\end{aligned}
\tag{3.35}
$$

where

$$
\mathcal{K} =
\begin{bmatrix} D_K & C_K \\ B_K & A_K \end{bmatrix},
\tag{3.36}
$$

and

$$
A_0 = \begin{bmatrix} A & 0 \\ 0 & 0 \end{bmatrix}, \quad
B_0 = \begin{bmatrix} B_w \\ 0 \end{bmatrix}, \quad
C_0 = \begin{bmatrix} C_z & 0 \end{bmatrix}, \quad
D_0 = D_{zw},
$$

$$J = \begin{bmatrix} B_u & 0 \\ 0 & I \end{bmatrix}, \quad F = \begin{bmatrix} D_{zu} & 0 \end{bmatrix}, \quad R = \begin{bmatrix} C_y & 0 \\ 0 & I \end{bmatrix}, \quad T = \begin{bmatrix} D_{yw} \\ 0 \end{bmatrix}.$$

For convenience, we define the following controller set

$$\mathcal{S} = \{\mathcal{K} \mid \mathcal{K} \text{ as defined in (3.36), with } A_K, B_K, C_K \text{ block diagonal}\}.$$

Note that with the introduction of the controller matrix $\mathcal{K}$, we have rewritten the problem of finding a dynamic output feedback controller, into the problem of finding a static output feedback controller as described in the previous section. The only difference is now that not the whole controller matrix $\mathcal{K}$ is used for structure optimization, but only the upper left part corresponding to D_k.

Since $D_K^{ij} \in \mathbb{R}^{q_{u,i} \times p_{y,i}}$, decentralization corresponds to zero matrices in the (i,j)-th block of the matrix D_K. To achieve this decentralized structure, we use the following pattern operator

$$\mathcal{Z}(\mathcal{D}_K, \infty) := \begin{bmatrix} \|D_K^{11}\|_\infty & \cdots & \|D_K^{1N}\|_\infty \\ \vdots & \ddots & \vdots \\ \|D_K^{N1}\|_\infty & \cdots & \|D_K^{NN}\|_\infty \end{bmatrix}.$$

Each element z_{ij} of $\mathcal{Z}(\mathcal{D}_K, \infty)$ represents one measurement entering the feedthrough matrix of the subcontroller (3.32), and $z_{ij} = 0$ if and only if $D_K^{ij} = 0$. We are no ready to state the decentralized dynamic output feedback problem.

3.5.1. Decentralized Control Problem – Dynamic Output Feedback

While for the unconstrained dynamic output feedback problem a convex formulation exists and the global optimum can be found, this is not the case if structural constraints are considered. Therefore, similar to the static output feedback case, we directly optimize the decentralized controller. This can be formulated in the following optimization problem

Problem 3.5 (Decentralized Dynamic Output Feedback Control Problem).
Given the linear system (3.1), determine a structured controller (3.33), such that

(i) *the closed-loop system $\Sigma_{cl}(\Sigma_K)$ in (3.34) is asymptotically stable and*

(ii) *the 0-norm of the pattern operator $\mathcal{Z}(D_K, \infty)$ is minimized subject to a given $\mathcal{H}_\infty$-performance constraint $\gamma > 0$, that is*

$$\min_{\Sigma_K \text{ stabilizing}} \|\mathcal{Z}(D_K, \infty)\|_0 \tag{3.37a}$$

$$\text{subject to} \quad \|\Sigma_{cl}(\Sigma_K)\|_\infty < \gamma. \tag{3.37b}$$

Again, it is also possible to directly consider $\|D_K\|_0$ instead of $\|\mathcal{Z}(D_K, \infty)\|_0$.

In the previously formulated optimization problem, we observe the same difficulties as in the state feedback case. Namely, the combinatorial nature of the objective function as well as the non-convex performance constraint. We will use similar approaches as in the state feedback case to derive a relaxed convex optimization problem.

3.5.2. Convex Relaxation – Dynamic Output Feedback

The convex relaxation of the objective function is equivalent to the relaxation presented in Section 3.3.5 and we can replace the non-convex 0-norm minimization by a weighted ℓ_1-minimization

$$\min_{\Sigma_K \text{ stabilizing}} \|\text{vec}(M \circ \mathcal{Z}(D_K, \infty))\|_1, \tag{3.38a}$$

$$\text{subject to } \|\Sigma_{cl}(\Sigma_K)\|_\infty < \gamma. \tag{3.38b}$$

Since we rewrote the closed loop such that we have a static output feedback problem in equation (3.35), we are able to adapt the system augmentation approach as presented in the previous section to this type of problem as well. Define the auxiliary variable

$$v = \mathcal{K}R\zeta + \mathcal{K}Tw \tag{3.39}$$

and consider $\zeta_s^T = \begin{bmatrix} \zeta^T & v^T \end{bmatrix}$ as an augmented variable, then we can write system (3.34) as a singular system similar to (3.24)

$$\Sigma_s : \quad \begin{bmatrix} E_s \dot{\zeta}_s \\ z \end{bmatrix} = \begin{bmatrix} A_s & B_s \\ C_s & D_s \end{bmatrix} \begin{bmatrix} \zeta_s \\ w \end{bmatrix}. \tag{3.40}$$

with

$$E_s = \begin{bmatrix} I & 0 \\ 0 & 0 \end{bmatrix}, \quad A_s = \begin{bmatrix} A_0 & J \\ \mathcal{K}R & -I \end{bmatrix}, \quad B_s = \begin{bmatrix} B_0 \\ \mathcal{K}T \end{bmatrix}, \quad C_s = \begin{bmatrix} C_0 & F \end{bmatrix}, \quad D_s = D_0.$$

With this coordinate transformation, we can apply Theorem 3.3 and Theorem 3.4 to the system given in equation (3.40). Again, the advantage of Theorem 3.4 lies in the fact that the structural constraints added to $\mathcal{K}$ can now be imposed on L explicitly. More specifically, we choose

$$L = \begin{bmatrix} D_L & C_L \\ B_L & A_L \end{bmatrix}$$

with partitioning compatible to $\mathcal{K}$, and there holds $\mathcal{Z}(D_K, 0) = \mathcal{Z}(D_L, 0)$.

The same conclusions hold as in the static output feedback case discussed in the previous section and we can formulate Algorithm 3.4 for decentralized dynamic output feedback control.

Due to the similar nature of Algorithm 3.4 and Algorithm 3.2 Remark 3.7 applies here, too. Additionally, to improve the convergence properties of Algorithm 3.4, Algorithm 3.3 should be applied first to improve the initial values for U_1 and V_1.

3.5.3. Discussion of Alternative Dynamic Output Feedback Controllers

The controller structure introduced in (3.32) has a special structure as the dynamic parts of the controller are completely decentralized and information exchange between the different subsystems is only allowed in the feed through matrix of the

---

Algorithm 3.4 Relaxed decentralized controller design algorithm (Dynamic output feedback)

---

1. Set $k = 1$ and choose initial values for U_1 and V_1. Choose a sufficiently small number $\nu > 0$ and set $\alpha > 0$.

2. For fixed U_k and V_k, solve the following convex optimization problem for the parameters $\Gamma = \{P_1 > 0, \text{ diagonal } P_2 > 0, L \in \mathcal{S}, \varepsilon_k\}$:

$$\inf_{\Gamma} \|\text{vec}(M_k \circ \mathcal{Z}(D_K, \infty))\|_1 + \alpha \varepsilon_k$$

subject to $\Pi(U_k, V_k) < \varepsilon_k I$.

3. If $\varepsilon_k < 0$, update U_{k+1} as $U_{k+1} = P_{2k}^{-1} L_k R$, $V_{k+1} = P_{2k}^{-1} L_k T$ and $m_{k+1}^{ij} = (|z_k^{ij}| + \nu)^{-1}$, where $\nu > 0$ ensures that the inverse is always well defined. Terminate on convergence or if $k = k_{\max}$ and go to Step 4. Otherwise set $k = k + 1$ and go to Step 2.
   If $\varepsilon_k > 0$, set U_k as $U_k = P_{2k}^{-1} L_k R$ and V_k as $V_k = P_{2k}^{-1} L_k T$. Go to Step 2.

4. Solve the following convex feasibility problem

$$\Pi(U_k, V_k) < 0$$

for the fixed controller structure obtained in Step 3. The controller is given as $\mathcal{K} = P_{2k}^{-1} L_k$.

---

controller. As stated before, this can be either seen as an extension to an existing decentralized control scheme or as an addition to static output feedback control. However, the most general dynamic output feedback controller has the following form

$$\begin{aligned}
\dot{x}_{K,i} &= \sum_{j=1}^{N} A_K^{ij} x_{K,j} + \sum_{j=1}^{N} B_K^{ij} y_j \\
u_i &= \sum_{j=1}^{N} C_K^{ij} x_{K,j} + \sum_{j=1}^{N} D_K^{ij} y_j,
\end{aligned} \tag{3.41}$$

with the transfer function

$$K^{ij}(s) = C_K^{ij}(sI - A_K^{ij})^{-1} B_K^{ij} + D_K^{ij}.$$

I.e. no controller matrix has a pre-specified structure and all controller matrices are part of the structure optimization. The complete controller is then given by

$$\Sigma_K : \quad = \begin{bmatrix} \dot{x}_K \\ u \end{bmatrix} = \begin{bmatrix} A_K & B_K \\ C_K & D_K \end{bmatrix} \begin{bmatrix} x_K \\ y \end{bmatrix}$$

Decentralization occurs now, if no link form subsystem j to subsystem i exists, i.e. if $K^{ij}(s) = 0$. This is only achieved if all matrices of one subcontroller are identically zero. Therefore, the following pattern operator is defined to represent the structure of the controller

$$
\mathcal{Z}\left(\begin{bmatrix} A_K & B_K \\ C_K & D_K \end{bmatrix}, \infty\right) := \begin{bmatrix} \left\| \begin{bmatrix} A_K^{11} & B_K^{11} \\ C_K^{11} & D_K^{11} \end{bmatrix} \right\|_\infty & \cdots & \left\| \begin{bmatrix} A_K^{1N} & B_K^{1N} \\ C_K^{1N} & D_K^{1N} \end{bmatrix} \right\|_\infty \\ \vdots & \ddots & \vdots \\ \left\| \begin{bmatrix} A_K^{N1} & B_K^{N1} \\ C_K^{N1} & D_K^{N1} \end{bmatrix} \right\|_\infty & \cdots & \left\| \begin{bmatrix} A_K^{NN} & B_K^{NN} \\ C_K^{NN} & D_K^{NN} \end{bmatrix} \right\|_\infty \end{bmatrix}.
$$

The approach for decentralized controller design presented in this thesis works for the modified controller structure above in the same way as for the structure in (3.32) by using the modified pattern operator for the weighted ℓ_1-minimization and dropping the structure constraint $L \in \mathcal{S}$. Therefore all theoretical contributions of this chapter are valid as well. However, we believe that this type of controller structure has several disadvantages from a practical point of view and was therefore not considered in the first place.

While the weighted ℓ_1-minimization is a powerful tool to achieve sparse matrices, it works better the smaller the blocks are that should be identically zero. In the modified pattern operator as introduced above, the blocks are much larger than in the originally considered pattern operator. Additionally, in a decentralized controller, the individual subcontrollers are spatially distributed. If we allow for the individual subcontroller to not only have local dynamics but also be influenced by the states of the other subcontrollers, information exchange increases significantly. In addition to the measurements, also the subcontroller states have to be transmitted. Alternatively, the complete dynamic matrix of the controller A_K has to be implemented at each subcontroller. In this case, each controller has significantly more states than the individual subsystem which is in general not a favorable option. As a consequence, we can see that a more general structure of the decentralized dynamic output feedback controller is possible but not favorable.

3.6. Application Example: Wide Area Control of Power Systems

The presented approach for decentralized controller design will be illustrated by designing a wide area controller for damping inter-area oscillations in a power system. Power systems describe large power grids in order to analyze stability and performance issues like frequency and voltage stability. The phase and frequency dynamics of power systems depend on the active power injection and consumption of generators, usually synchronous generators, and loads. Only for illustration, we think of these dynamics as coupled, weakly damped oscillators where the individual oscillators represent the rotating mass of the synchronous generators

$$
M\ddot{\theta} + D\dot{\theta} + L\theta = 0, \tag{3.42}
$$

where θ and $\dot{\theta}$ are the rotor angles and frequencies, M and D are the diagonal matrices of inertia and damping coefficients and L is the Laplacian (or admittance) matrix (see e.g. Kundur (1994) for more details on modeling of power systems). A more detailed model will be used later on. Yet, this model (3.42) is used here to explain the problem setup. Due to the weak damping of the oscillators, these power systems often exhibit several modes oscillating between these generators. The most critical of these oscillations are inter-area oscillations where groups of generators oscillate against each other. State of the art methods use power system stabilizers (PSS) that adjust the power supply of the individual generators based on local frequency measurements, i.e. PSS are decentralized controllers for the power grid. The PSS are tuned such that the damping of the individual oscillators is increased for certain frequency intervals where the critical local and inter-area oscillations occur. Sometimes, it is not possible to stabilize inter-area modes by PSS or, when decentralized control is stabilizing, the performance is poor (Seethalekshmi et al., 2008). In these cases, wide area controllers could be used, i.e. locally implemented controllers that use remote measurement signals to improve the performance of the overall network. Especially flexible AC transmission systems (FACTS) could be controlled using these remote measurements to effectively damp inter-area oscillations (Seethalekshmi et al., 2008). Due to technological advances on FACTS and communication systems, wide-area control is nowadays technically feasible. However, it is still far from being a standard approach in power systems, also due to the complexity of designing such controllers. In addition, communication is often expensive, since measurements have to be transmitted over large distances, e.g. continental Europe, and a centralized control of such a large system over such distances is not possible. The idea behind the example is to reduced the investment cost for the additional FACTS and invest instead in a reliable high-speed communication line between individual components and measurement points. Clearly, this investment pays off if only very few communication lines are needed. Therefore, sparsity of the controller is very critical for this application.

We consider the 3-area, 6-machine power system introduced in Taranto et al. (1994) and shown in Fig. 3.2 to illustrate the presented approach for decentralized controller design. The circles in the figure depict the six generators, while the 14 buses are numbered by #i, $i = 1, \ldots, 14$. The considered loads are given as L_j, $j = 3, 4, 6, 9, 10$. We consider here a much more detailed model than the one introduced in (3.42): The six generators are modeled by eight state variables each: six states describing the electromechanical dynamics of the generator and two states describing the excitation system of the synchronous generator including the excitation controller. We design the wide area controller for a power system model that is linearized around its steady state and consists of 48 states in total. The linearized model is generated using the Power Systems Toolbox (Chow and Cheung, 1992). The dynamic state variable x of the power system model is given by $x = [\theta^T, \dot{\theta}^T, x_{\text{rem}}^T]$, where $\theta, \dot{\theta} \in \mathbb{R}^6$ are the rotor angles and frequencies of the six synchronous generators and $x_{\text{rem}} \in \mathbb{R}^{36}$ are the remaining state variables corresponding to electrical and controller dynamics. The power system is stable and has five poorly damped modes of oscillations. The two low-frequency modes are the inter-area modes, which we want to damp using a wide-area controller (see Table 3.1). The other three modes are the local modes of machine oscillations within the areas. The mode shapes of the inter-area modes indicate that there are three coherent groups: machines 1 and 2 form the first coherent area C, machines 3 and

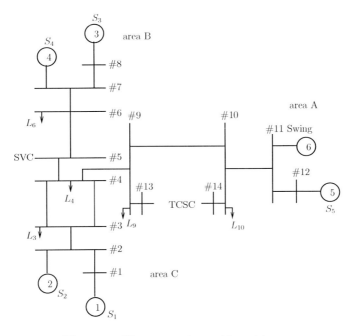

Figure 3.2.: Three-area, six-machine system.

4 form the second coherent area B, and machines 5 and 6 form the third coherent area A. The first inter-area mode (mode 1) consists of the machines of area B oscillating against the machines of areas A and C, and the second inter-area mode (mode 2) consists of the machines of area A oscillating against the machines of area C. Table 3.1 denotes the eigenvalues of these modes. Figure 3.3 illustrates the elements of the eigenvectors of these modes that show which areas are oscillating against each other. Figure 3.6(a) and Figure 3.7(a) show the frequency time series of the open loop when the initial values are aligned with the respective inter area mode. In a first step, a nominal decentralized controller is designed for this system. The aim of this controller is to either damp or disorder the inter-area modes. In a second step this controller is analyzed in terms of robustness in face of tie-line strength variations which occur regularly in power systems.

3.6.1. Nominal Controller Design

To damp the inter-area modes, the system is equipped with additional actuators, namely a Thyristor Controlled Series Capacitor (TCSC) and a Static VAR compensator (SVC), i.e. we have two control inputs to damp the inter-area modes. Both devices belong to the FACTS family. As suggested in Taranto et al. (1994) and

Table 3.1.: Inter-area modes of the 3-area, 6-machine system.

mode no.	eigenvalue pair	frequency [Hz]	damping ratio [%]	coherent groups
1	$-0.0111 \pm 3.6773i$	0.5853	0.30	3, 4 vs. 1, 2, 5, 6
2	$-0.2638 \pm 7.3397i$	1.1682	3.59	1, 2 vs. 5, 6

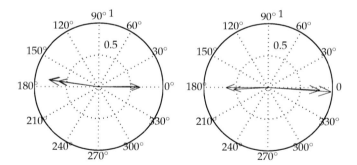

Figure 3.3.: Illustration of the elements of the eigenvectors of the inter-area modes of the 3-area, 6-machine system, left mode 1, right mode 2. Each arrow represents a generator that swings with a certain amplitude and phase. Inter-area modes are characterized by two or more arrows with a phase shift of about 180° against each other.

Taranto and Falcao (1998) the SVC is placed at the midpoint between area B and area A and C, which is roughly at bus 5. The TCSC is located in the power transfer path between area A and C between bus 13 and 14 (see Figure 3.2). Both SVC and TCSC bring additional states into the system, such that the overall power system including these FACTS has 51 states. We assume that disturbances act in the same way as the controlled inputs, i.e. $B_w = B_u$. This is motivated by the fact, that we want to reject communication noise in the wide-area implementation. To damp the oscillations of the system using the FACTS, we penalize frequency violations and angular differences, as well as the input signals, i.e

$$z_1 = w_1 E\theta$$
$$z_2 = w_2 I\dot{\theta}$$
$$z_3 = w_3 I u,$$

where E is the incidence matrix that captures the angular differences. The weighting functions w_1, w_2, and w_3 are chosen to be static gains. The choice of performance outputs is motivated by the idea that an *ideal* power system without inter-area oscillations is characterized by homogenous interactions and uniform inertia coefficients (Kundur, 1994). In an ideal power system, the network is homogenous and composed of identical generators, i.e. all generators swing in synchrony. The performance output z_1 penalizes the difference between two different angles $\theta_i - \theta_j$ for any combination of i and j. The performance output z_2 penalizes the angle speed deviation $\dot{\theta}_i$ and the performance output z_3 penalizes the control input u. These performance outputs will lead to an improved average closed-loop performance with all inter-area modes either damped or distorted. The performance outputs do not penalize steady state deviations in the generator d/q axis voltages or the state variables of the excitation system included in the A-matrix. A similar setup for damping of inter-area modes can also be found in Dörfler et al. (2013); yet in terms of an $\mathcal{H}_2$-optimization instead of an $\mathcal{H}_\infty$-optimization.

The *centralized* controller $\hat{\mathcal{K}}$ using the above performance outputs is the result of a convex optimization problem. It uses all local and remote signals for state feedback in the SVR and TCSC (i.e. $\|\hat{\mathcal{K}}\|_0 = 102$). The achieved closed loop performance is $\|\Sigma(\hat{\mathcal{K}})\|_\infty = 1.0002$ and the damping factor of the least damped mode is increased from 0.3% to 12.42%. We now want to find decentralized controllers $\mathcal{K}$ that use less remote signals to achieve (almost) the same performance as the centralized controller using Algorithm 3.1. The resulting optimization problem has 5357 decision variables and was implemented in Matlab, using SeDuMi (Sturm, 1999) and Yalmip (Löfberg, 2004). We compute in total 40 different controllers: 30 for logarithmically spaced performance degradation of $[0.001\%, 30.62\%]$ compared to the centralized controller and another 10 for $[30.62\%, 1000\%]$. We use a homotopy approach for the initial weighting matrix M_1 of the ℓ_1-minimization starting with the element-wise inverse of the centralized controller and using the previously designed decentralized controller for the next degradation level. For each controller, the iteration for the ℓ_1-weight update was set to 4 (i.e. $k_{\max} = 4$ in Algorithm 3.1). Additionally, an homotopy approach is used for U_1, starting with $U_1 = [0 \ \hat{\mathcal{K}}]$ for the first degradation level followed by the newly computed decentralized controller $U_1 = [0 \ \hat{\mathcal{K}}]$ for the next degradation level. Due to this choice of U_1, $\Pi(U_1) < 0$ is always satisfied.

To get more insight which of the system states contribute most to the damping, we use $\|M \circ \mathcal{K}\|_1$ as objective function.

Figure 3.4 and Figure 3.5 show the results of the optimization. In Figure 3.4, it can be seen, that the sparsity of the controller increases for higher allowed performance degradation, i.e. $\|\mathcal{K}\|_0$ decreases. Note that the dashed line at $\|\mathcal{K}\|_0 = 102$ corresponds to the centralized controller and $\gamma = 0$ which cannot be depicted in the logarithmic scale. By allowing a performance degradation of only $\gamma = 1.0002 \cdot 10^{-5}$, we can decrease the number of non-zero elements to $\|\mathcal{K}\|_0 = 44$ and for the maximally allowed performance degradation only two states have to be transmitted. Figure 3.5 shows exemplarily the non-zero controller elements for five different levels of allowed performance degradation γ. It can be seen that, for increasing sparsity, the SVC uses preferred information of generator 2, 3 and 6 as well as its local information. These generators belong to the different coherent groups A (generator 2), B (generator 3) and C (generator 6) which exhibit inter-area mode 1 in the open loop. I.e., to damp this inter-area mode, the SVC uses informations of the generators that oscillate against each other. The TCSC uses information of generator 1 (coherent group C) and generator 5 (coherent group A) as well as local information. Again, the TCSC uses information of the two different coherent groups to damp the second inter area mode. For the closed loop with $\|\mathcal{K}\|_0 = 9$ the least damped mode has still a damping factor of 11.99% which is an increase of factor 41 compared to the damping ratio of 0.3% of the open loop. This is a performance degradation in terms of the damping factor of only 3.5% compared to the centralized controller, while we can omit more than 90% of the originally required feedback. Finally, we can observe, that the choice of our performance outputs encourages the use of generator angles and speed as feedback additionally to the local feedback. Both, angles and speed, are readily available in most cases since they are also used for the local controllers (i.e. PSS). Figure 3.6 and Figure 3.7 show the frequency time series of the open and closed loop when the initial values are aligned with the inter-area modes of the open loop. It can be seen that the damping is increased in all cases and that the frequencies do not swing against each other any more but are much more aligned. Hence, the controller achieves its goal.

The proposed algorithm is able to identify the most significant signals needed to be transmitted in a wide-area control framework to damp inter-area oscillations. Moreover, we are able to almost reproduce the performance of the centralized controller whereas the number of transmitted signals is significantly reduced.

3.6.2. Robustness Analysis of the Nominal Controller

To investigate the robustness properties of the designed controllers, we analyze the performance of the decentralized controllers in face of different operation conditions of the power system. As suggested in Taranto et al. (1997), we consider operating conditions with tie-line strength variations Z_{eq5-6} between buses 5 and 6, and Z_{eq9-10}, between buses 9 and 10. Note that this are the critical lines for the inter-area modes we want to damp. The line between buses 5 and 6 connects area B and C and the line between buses 9 and 10 connects area A with B and C. Two scenarios, each with three operating conditions, are analyzed. Scenario 1 represents

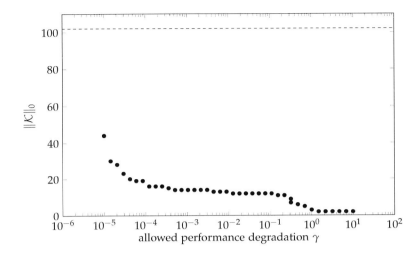

Figure 3.4.: Number of non-zero elements in the controller matrix for different allowed performance degradation levels γ. The dashed horizontal line $\|\hat{\mathcal{K}}\|_0 = 102$ corresponds to the centralized controller ($\gamma = 0$).

the variation of Z_{eq5-6} in the SVC transmission path, and scenario 2 represents the variation of $Z_{eq,9-10}$ in the TCSC transmission path. The operating conditions represent the weak, median and stiff transmission strength configurations, with the power transfer between the areas remaining the same. The median configuration corresponds to the nominal system for which the controller was designed and is the same in both cases. The damping factor of the least damped mode for different levels of decentralization are summarized in Table 3.2 and Table 3.3.

The damping factor of the least damped mode is increased in all cases compared to the open loop, except for the centralized controller with tie-line strength variations Z_{eq9-10}. Note, that the damping factor of the least damped mode increases in the weak and stiff operation conditions compared to the median operation condition and we achieve acceptable performance in all cases. Additionally one can

Table 3.2.: Damping factor of the least damped mode for tie-line strength variation Z_{eq5-6} between buses 5 and 6 for different decentralization levels.

System	Z_{eq5-6},pu	open loop	$\|\mathcal{K}\|_0$				
			102	44	23	9	2
weak	$0.095 + j0.95$	5.19%	6.04%	6.54%	6.42%	5.35%	5.23%
median	$0.080 + j0.80$	0.3%	12.42%	12.04%	12.10%	12.28%	8.04%
stiff	$0.050 + j0.50$	5.08%	5.96%	6.49%	6.83%	5.37%	5.19%

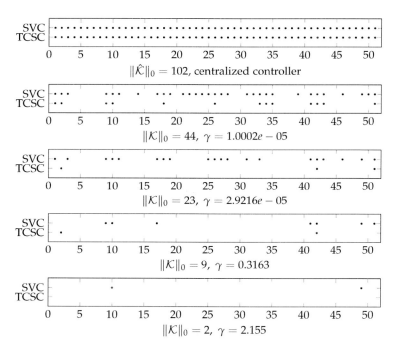

Figure 3.5.: Non-zero elements in the controller matrix for different allowed performance degradation levels γ.

Table 3.3.: Damping factor of the least damped mode for tie-line strength variation Z_{eq9-10} between buses 9 and 10 for different decentralization levels.

System	Z_{eq9-10},pu	open loop	$\|\mathcal{K}\|_0$				
			102	44	23	9	2
weak	$0.280 + j2.80$	4.43%	14.75%	13.72%	10.95%	5.98%	5.05%
median	$0.040 + j0.40$	0.3%	12.42%	12.04%	12.10%	12.28%	8.04%
stiff	$0.010 + j0.10$	4.71%	4.18%	5.20%	6.22%	6.29%	5.59%

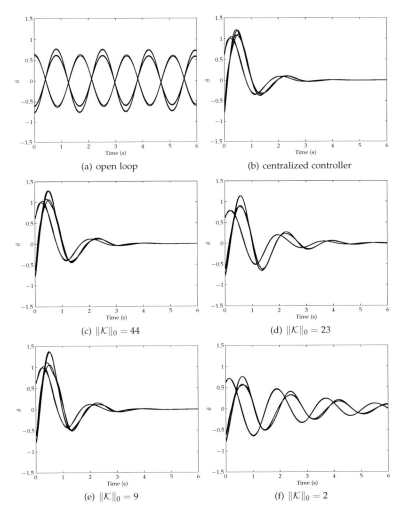

Figure 3.6.: Frequencies in open and in closed loop for different decentralization levels. The initial conditions are aligned with the eigenvector corresponding to the open-loop area mode 1.

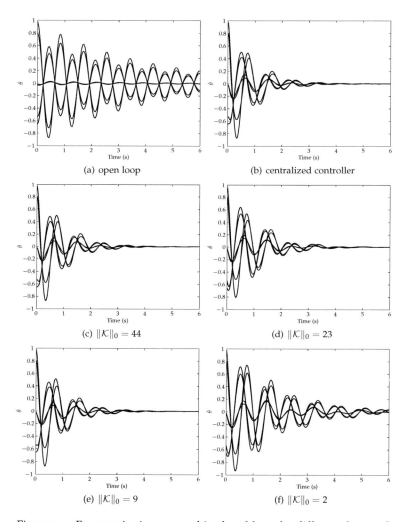

Figure 3.7.: Frequencies in open and in closed loop for different decentralization levels. The initial conditions are aligned with the eigenvector corresponding to the open-loop area mode 2.

see, that the damping factor of the least damped mode is not a monotonically decreasing function in $\|\mathcal{K}\|_0$. This is due to the fact, that the chosen performance output tries to improve the overall performance of the system, not just the least damped mode. To further improve the robustness of the designed controller in face of model uncertainties, robustness constraints could be added directly to the optimization problem.

3.7. Summary

In this chapter, a new design method for $\mathcal{H}_\infty$ decentralized control of multivariabel interconnected subsystems has been presented. These systems are omnipresent in many application fields, e.g. distributed power generation networks, highly coupled chemical plants or chemical reaction networks. We have formulated the decentralized controller design problem by minimizing the 0-norm of a pattern operator representing the structure of the controller to optimize the controller topology. In contrast to existing design methods for decentralized control, the structure of the decentralized controller and the controller dynamics are designed *jointly*. The minimization of the pattern operator is the objective function of the optimization problem resulting in sparse controllers while the constraints guarantee a specified $\mathcal{H}_\infty$-performance. Different approaches where presented, to design state feedback, static and dynamic output feedback controllers.

For the resulting combinatorial optimization problems, computationally tractable convex relaxations have been provided. More specifically, by means of a system augmentation approach and a re-weighted ℓ_1-relaxation, iterative algorithms were developed to deal with the relaxed decentralized control problem. Furthermore, we have shown how to choose the initial value for the iteration to improve feasibility. Especially in the state feedback case, we utilize that the optimal solution to the centralized control problem is known. The centralized controller is used to initialize the iterative algorithm to make convergence to the global optimum more likely. Furthermore, the performance degradation due to decentralization is exactly known. This is not possible in the output feedback case, because the optimal centralized controller cannot be computed using convex optimization techniques. We therefore apply approaches known from static output feedback to solve this problem. The proposed algorithm can be easily adapted to incorporate additional performance constraints, e.g. robustness considerations as long as they can be formulated as convex problems.

The proposed design method was applied to design wide area control to damp inter-area oscillations in a power systems example taken from literature. For the considered 3-area, 6-machine system, the algorithm could identify the most important feedback signals such that the damping of the system is increased. The performance of the centralized controller could be achieved using only very few communication links and both inter-area modes were either damped or distorted. A robustness analysis shows that the designed controllers can also damp inter-area oscillations in face of lie-line strength variations on the critical lines.

4. Sparse Topology Design for Dynamical Networks

4.1. Overview

In this chapter we focus on the design of networked dynamical systems. Instead of designing sparse decentralized controllers as in the previous section, we want to design the *network topology* i.e., the plant architecture. We will discuss two cases: design of the open loop system and design of the closed loop system. When designing networked dynamical systems, the engineer often has (some) freedom in the placement of sensors and actuators or in the coupling between the individual agents. We will review how different choices of network topology influence the performance of the open and closed loop system. In the following, two specific problems are studied in more detail: Namely the consensus network as a canonical model for networked dynamical systems. Here we design the closed loop system. And second, relative sensing networks (RSN) as a special class of networks with more complex dynamics where we design the open loop system.

The consensus model can be seen as a canonical form of networked dynamical systems. The model is most often presented as a collection of single integrators interacting over a communication graph. To study these type of systems, graph theory is the right mathematical tool. Perhaps the most well known result in this field is the relationship of the algebraic connectivity of a graph, sometimes referred to as the Fiedler eigenvalue (Fiedler, 1973), to the convergence rate of the dynamical system. Therefore, the dynamical behavior of the system shows a strong connection with the underlying graph topology (Fax and Murray, 2004; Mesbahi and Egerstedt, 2010). Consensus models are not only of interest because to their theoretic impact and simplicity but appear in many application areas ranging from optimization and sensor fusion to problems in formation control and distributed control and estimation (Fax and Murray, 2004; Giordano et al., 2011; Nedic and Ozdaglar, 2009; Olfati-Saber and Shamma, 2005; Olfati-Saber, 2005).

Even though the analysis of consensus systems has matured, there are only very few results that characterize the performance of a consensus system beyond its convergence rate or Fiedler eigenvalue (Dai and Mesbahi, 2011; Kim and Mesbahi, 2005). Nevertheless, the importance of this property has been extended in many directions, including convergence analysis for random graphs (Hatano and Mesbahi, 2005) and graphs with communication delays and switching topologies (Olfati-Saber and Murray, 2004). Various modifications to the consensus model have also led to more general system theoretic notions such as controllability and observability (Nabi-Abdolyousefi and Mesbahi, 2011; Rahmani et al., 2008) and input-output properties (Briegel et al., 2011; Tonetti and Murray, 2009; Yoon et al.,

2011) of dynamical networks. An important extension of these results is therefore a more general notion of system performance e.g., classical input-output behavior known from robust control theory and their relation to the network topology. A first step towards this direction is to include exogenous inputs in the form of noises and disturbances into the consensus model. Such models have been considered in Xiao et al. (2007); Das et al. (2010) and Hatano et al. (2005) where noises were introduced in either the process or measurement of the consensus model. Noises in the consensus model lead to a drift in the variance of the agreement value, and attempts to compensate for this include the design of edge weights (Xiao et al., 2007), or the introduction of a time-varying gain on the control to effectively reject the noise asymptotically (Das et al., 2010). When the consensus model is affected by noise, one can also consider the $\mathcal{H}_2$ performance. It is a measure of the noise's effect on the asymptotic deviation of each node's state from the consensus state (Scardovi et al., 2010). In Lin et al. (2011), the leader selection problem for consensus networks was considered within an $\mathcal{H}_2$ framework, while the work in Patterson and Bamieh (2011) examined the $\mathcal{H}_2$ performance of a networked system in fractal graphs.

We also consider consensus networks with exogenous inputs and performance outputs in an $\mathcal{H}_2$-framework. Our main focus is the *design* of these networks. A fundamental challenge for the design of consensus networks is the computational complexity of solving combinatorial problems. A common approach to this problem is to consider optimization over weighted graphs or other convex relaxations (Boyd, 1998; Boyd and Ghosh, 2006; Xiao et al., 2007). However, we emphasize a distinction between spectral properties of the graph, i.e. the eigenvalues of the Laplacian matrix, and combinatorial properties of a graph such as path lengths and cycles. When working with spectral properties, an additional layer of abstraction is introduced and direct graph properties such as edge costs and distances are more difficult to handle. A thorough treatment in this direction was recently given in Zelazo and Mesbahi (2011a) via the introduction of the *Edge Laplacian* and its corresponding *edge agreement problem*. The edge Laplacian is a variant of the graph Laplacian that provides a better understanding of how spanning trees and cycles affect certain algebraic properties of a graph. When the consensus model is analyzed using this construction, clear graph theoretic interpretations of the $\mathcal{H}_2$ norm of the system can be derived (Zelazo and Mesbahi, 2011a).

It is well known that the addition of cycles in a graph will increase its algebraic connectivity (Godsil and Royle, 2009), i.e. the algebraic connectivity of a undirected graph is a non-decreasing function of the number of edges. In Zelazo and Mesbahi (2011a) it was additionally shown that the $\mathcal{H}_2$ performance of consensus networks is lower bounded by the complete graph, and upper bounded by any choice of a spanning tree. To the best of the author's knowledge, there are no similar results for other performance metrics known. The analytical results from Zelazo and Mesbahi (2011a) are used in this thesis to formulate a synthesis problem for consensus networks. The problem considers the task of adding a fixed number of edges to an existing consensus network such that the performance improvement is maximal. A first approach to this problem leads to a mixed-integer program, generally considered a hard problem to solve due to its combinatorial nature. We will then reformulate the mixed-integer program into a weighted ℓ_1-optimization problem and present a numerically efficient solution. We also highlight how the weighting mechanism used in the ℓ_1-formulation provides an important tuning parameter for design. This formulation also allows to consider additional performance criteria including maximizing the algebraic connectivity of the graph. The provided solutions to this problem are visualized via simulation examples. Results for the design of consensus networks are based on Zelazo et al. (2012, Section IV and V) and Zelazo et al. (2013, Section 5 and 6).

The second type of models we consider are relative sensing networks (RSN). They consist of single agents that rely on relative sensing to achieve a common mission. In their most general form, relative sensing networks are a collection of autonomous systems that use relative output information to achieve this mission. In contrast to the consensus problem, the individual agents can be of higher order and have more complex dynamics. While in the consensus model the individual agents are coupled at state level, this is not the case for relative sensing networks. Here, the single systems are spatially distributed and have an inherently distributed sensing architecture. Therefore, the underlying sensing topology couples the agents at their outputs and therefore introduces an *implicit* network. The information exchange between different agents in an RSN describes the underlying connection topology. We will focus on how to design this underlying connection topology such that performance goals are fulfilled. This means that in the considered problem setup, we design the *open loop* system.

The ability of a single agent using available sensors to measure state information of an entire network can be limited by spatial constraints such as orientation, range and power requirements (Bauer et al., 1999; Corazzini et al., 1997; Purcell et al., 1998). Applications for distributed sensor networks relying on relative sensing range from environmental surveillance, modeling and localization to collaborative information processing (Akyildiz et al., 2002; Barooah and da Silva, 2006; Khan et al., 2009; Olfati-Saber, 2005). Additionally, such systems are relevant in formation flying applications where distributed sensing is employed to measure inter-agent distances (Mesbahi and Hadaegh, 1999; Smith and Hadaegh, 2005).

The analysis and synthesis of relative sensing networks was recently considered in Zelazo and Mesbahi (2011b) with respect to $\mathcal{H}_2$ and $\mathcal{H}_\infty$-performance. Strong results exist for unweighted homogenous graphs relating network properties to the $\mathcal{H}_\infty$-performance of the network. When dealing with heterogenous agents in a network, this is modeled by *node and edge weighted* graphs. The dynamics of the single agent influences the performance of the network and the links between the agents differ in importance or fidelity. The agent dynamics can be interpreted as a node weight on the graph, whereas the link dynamics (or static gain) is an edge weight. Therefore, one very often deals with weighted graphs in reality. However, there exist only very few analytical results for the design of node and edge weighted graphs.

Another important topic when dealing with networks of dynamical systems is robustness to uncertainties. The network should retain certain properties such as algebraic connectivity or $\mathcal{H}_\infty$-performance even though parts of the network dynamics are unknown. Robustness of network topologies is also considered in the area of survivable network design (Ellison et al., 1997; Kerivin and Mahjoub, 2005), but there, control theoretic aspects are often not taken into account. We contribute to the topology design for networks with heterogenous agents modeled by node and edge weighted graphs. We will first focus on the nominal case, where the complete dynamics of the system are known. In a second step we will consider uncertainties in the network topology. In both cases it is our goal to design *sparse* relative sensing networks, i.e. networks that fulfill required properties with as few links as possible and are robust against uncertainties. The presented design algorithms can be reformulated as convex optimization problems and can be solved efficiently using standard software. Results for the design of relative sensing networks are based on Schuler et al. (2012b, 2013c).

In this chapter, we use graph theory and especially graph Laplacians as a tool to *design* the topology of a network. A brief introduction into graph theory with special emphasis on the tools used in this thesis is given in Appendix C. A in-depth treatment of graph theory can for example be found in Godsil and Royle (2009).

4.2. Review of Models for Consensus and Relative Sensing Networks

This section gives a short review of models for consensus and relative sensing networks from the literature on which be base the presented design methods. First the consensus model is described followed by the model for relative sensing networks.

4.2.1. Consensus Networks

The standard consensus model is based on a collection of n single integrators that exchange relative state information over a communication graph. The model is usually presented as an autonomous system with no noises or disturbances (Mesbahi and Egerstedt, 2010),

$$\dot{x}(t) = -L(\mathcal{G})x(t). \tag{4.1}$$

Here, the vector $x(t) \in \mathbb{R}^n$ is the stacked state vector of the single agents state $x_i(t) \in \mathbb{R}$, and $L(\mathcal{G})$ is the Laplacian matrix of the graph $\mathcal{G}$.

A two-port interpretation of the consensus model provides a framework for considering the presence of exogenous inputs, such as reference signals and noises entering the measurement and process. We use the setup as presented in Zelazo and Mesbahi (2011a). The inputs are considered to be white Gaussian noise of unit covariance. These restrictions are for ease of notation only. This representation is pictured in Figure 4.1, and is described as

$$\begin{bmatrix} \dot{x}(t) \\ z(t) \\ y(t) \end{bmatrix} = \begin{bmatrix} 0 & I & 0 & I \\ E(\mathcal{G})^T & 0 & 0 & 0 \\ E(\mathcal{G})^T & 0 & I & 0 \end{bmatrix} \begin{bmatrix} x(t) \\ w(t) \\ v(t) \\ u(t) \end{bmatrix} \tag{4.2}$$

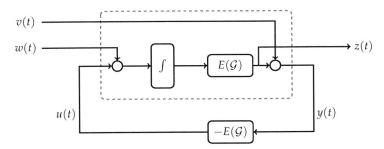

Figure 4.1.: Consensus as a 2-port feedback configuration.

with the control input defined as $u(t) = -E(\mathcal{G})y(t)$. The closed-loop system is the given by

$$\Sigma(\mathcal{G}): \quad \begin{bmatrix} \dot{x}(t) \\ z(t) \end{bmatrix} = \begin{bmatrix} -L(\mathcal{G}) & I & -E(\mathcal{G}) \\ E(\mathcal{G})^T & 0 & 0 \end{bmatrix} \begin{bmatrix} x(t) \\ w(t) \\ v(t) \end{bmatrix}. \tag{4.3}$$

The two-port representation of the consensus model is meant to illustrate the underlying mechanism of the dynamics; it transparently demonstrates how disturbances can enter into the model and also shows the distributed nature of the dynamics. This more general model is not a minimal realization of the system. Indeed, the system has an uncontrollable and unobservable mode in the direction of the $\mathbb{1}$ vector (Zelazo and Mesbahi, 2011a, 2010). Furthermore, due to the eigenvalue at the origin of the state matrix, certain system norms are not meaningful (i.e. the $\mathcal{H}_2$-norm of the system (4.3) is unbounded). As discussed in Zelazo and Mesbahi (2011a), a minimal realization of the system can be expressed via a coordinate transformation using the essential edge Laplacian matrix L_e (C.5),

$$\Sigma_e(\mathcal{G}): \quad \begin{bmatrix} \dot{x}_\tau(t) \\ z(t) \end{bmatrix} = \begin{bmatrix} -L_e(\mathcal{T})R_{\mathcal{T},\mathcal{C}}R_{\mathcal{T},\mathcal{C}}^T & E(\mathcal{T})^T & -L_e(\mathcal{T})R_{\mathcal{T},\mathcal{C}} \\ I & 0 & 0 \end{bmatrix} \begin{bmatrix} x_\tau(t) \\ w(t) \\ v(t) \end{bmatrix}, \tag{4.4}$$

with $R_{(\mathcal{T},\mathcal{C})} = \begin{bmatrix} I & T_{(\mathcal{T},\mathcal{C})} \end{bmatrix}$ and $T_{(\mathcal{T},\mathcal{C})}$ as defined in (C.3). Thus, (4.4) represents the dynamics for only the controllable and observable modes of the system. Here, the transformed state vector $x_\tau(t) \in \mathbb{R}^{|\mathcal{E}_\tau|}$ can be interpreted as a state associated with the edges of the spanning tree $\mathcal{T}$. The system (4.4) is referred to as the *edge agreement problem*. An added benefit of this system is that the state matrix is now Hurwitz; i.e., all eigenvalues are in the open left-half of the complex plain. In this work we consider $\Sigma_e(\mathcal{T})$ as a *skeletal system* for the complete consensus network. In this regard, we only observe the states along the tree, $x_\tau(t)$, as the controlled variable in the edge agreement problem. As opposed to considering $R_{(\mathcal{T},\mathcal{C})}^T x(t)$ as the controlled variable.

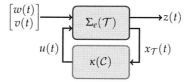

Figure 4.2.: Cycles as feedback mechanism.

As shown in Zelazo et al. (2013), the cycles in the graph can be viewed as an internal feedback mechanism for the minimal system $\Sigma_e(\mathcal{G})$. To visualize this idea, consider the following dynamic system over a spanning tree $\mathcal{T}$,

$$\Sigma_e(\mathcal{T}): \quad \dot{x}_\tau(t) = \begin{bmatrix} -L_e(\mathcal{T}) & E(\mathcal{T})^T & -L_e(\mathcal{T})R_{\mathcal{T},\mathcal{C}} & L_e(\mathcal{T}) \end{bmatrix} \begin{bmatrix} x_\tau(t) \\ w(t) \\ v(t) \\ u(t) \end{bmatrix}$$

with a state-feedback control $\kappa(\mathcal{C}) : u(t) = -T_{(\mathcal{T},\mathcal{C})}T_{(\mathcal{T},\mathcal{C})}^T x_\tau(t)$. This is also shown in Figure 4.2.

The interpretation of cycles as feedback leads to a better understanding of their role in the consensus problem Indeed, when cast as a feedback problem as in Figure 4.2, one can try to use cycles reduce the sensitivity of the system to external disturbances. The two-port feedback paradigm in control systems provides a powerful framework for formulating and solving problems related to *optimal controller design* (Dullerud and Paganini, 2000). Therefore, based on the interpretation of cycles as feedback, one can now attempt to relate the problem of *optimal design of graphs* to the problem of *optimal synthesis of feedback controllers*. This will be further exploited in Chapter 4.3, where we show a design procedure for consensus models.

4.2.2. Relative Sensing Networks

We consider a model for relative sensing networks as described in (Zelazo and Mesbahi, 2011b). It consists of a group of g linear time-invariant dynamical systems (agents)

$$\Sigma_i : \quad \begin{bmatrix} \dot{x}_i(t) \\ y_i(t) \end{bmatrix} = \begin{bmatrix} A_i & B_i \\ C_i & 0 \end{bmatrix} \begin{bmatrix} x_i(t) \\ w_i(t) \end{bmatrix} \tag{4.5}$$

where each agent is indexed by the script i. Here, $x_i(t) \in \mathbb{R}^{n_i}$ represents the state, $w_i(t) \in \mathbb{R}^{q_{wi}}$ the exogenous input and $y_i(t) \in \mathbb{R}^{p_{yi}}$ the measured output. The transfer-function representation of Σ_i is denoted as H_i with

$$H_i := C_i(sI - A_i)^{-1}B_i. \tag{4.6}$$

We assume a minimal realization for each agent and compatible output for all agents, e.g. system outputs will correspond to the same physical quantity. In the homogeneous case, it is assumed that each agent in the RSN possesses the same dynamics and is described by the same state-space realization (e.g. $\Sigma_i = \Sigma_j$ for

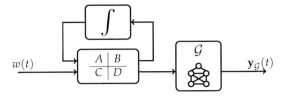

Figure 4.3.: Global RSN layer block diagram; the feedback connection represents an upper fractional transformation Dullerud and Paganini (2000).

all i, j). It should be noted that in a heterogeneous RSN the dimension of each agent can be different. The parallel interconnection of all agents can be expressed by a concatenation of the corresponding system states, inputs, and outputs, and through the block diagonal aggregation of each agent's state-space matrices.

The sensed output of the RSN is the vector $y_{\mathcal{G}}(t)$ containing relative output information of each agent and its neighbors and is motivated by the relative sensing problem discussed in Section 4.1. The incidence matrix of a graph naturally captures state differences and will be the algebraic construct used to define the relative outputs of RSNs. In this paper, we will especially consider that the relative output of the RSN corresponds to a relative position' measurement between all agents as

$$y_{\mathcal{G}}(t) = (WE(K_g)^T \otimes I)y(t). \tag{4.7}$$

Here K_g is the complete graph over the g agents and the topology is defined by the diagonal weighting matrix $W = \mathrm{diag}(w_1, \ldots, w_{|\mathcal{E}|})$, with $w_i \in \mathbb{R}_0^+$ and the node set given as $V = \{1, \ldots, g\}$. Note that edge i does not exist if and only if $w_i = 0$. Furthermore, we denote the vector containing all weights as $w = [w_1, \ldots, w_{|\mathcal{E}|}]^T$. The weights w_i in this setup can be seen as the gains of the sensor used to sense the relative state. They might be used to capture the fidelity of a relative measurement. The global layer is visualized in the block diagram shown in Figure 4.3.

Using the above notations, we can express the heterogeneous RSN in a compact form

$$\Sigma_{het}(\mathcal{G}) : \quad \begin{bmatrix} \dot{x}(t) \\ y(t) \\ y_{\mathcal{G}}(t) \end{bmatrix} = \begin{bmatrix} A & B & 0 \\ C & 0 & 0 \\ 0 & 0 & WE(K_g)^T \otimes I \end{bmatrix} \begin{bmatrix} x(t) \\ w(t) \\ y(t) \end{bmatrix} \tag{4.8}$$

The homogeneous RSN, $\Sigma_{hom}(\mathcal{G})$, can be expressed using Kronecker products, for example $A = I_g \otimes A_{him}$. The transfer function representation of Σ_{het} is denoted as $\hat{\Sigma}_{het}$ and is defined as in (4.6). The homogeneous system $\hat{\Sigma}_{hom}$ can also be written using Kronecker products in a similar manner as described above. We denote $T_{het}^{w \mapsto \mathcal{G}}$ and $T_{hom}^{w \mapsto \mathcal{G}}$ as the map from exogenous inputs $w(t)$ to the RSN sensed output $y_{\mathcal{G}}(t)$ for homogeneous and heterogeneous systems respectively, e.g. $T_{het}^{w \mapsto \mathcal{G}} = (WE(K_g)^T \otimes I_q)H$ and $T_{hom}^{w \mapsto \mathcal{G}} = WE(K_g)^T \otimes H(s)$.

In Chapter 4.4 we will focus on the design of relative sensing networks with respect to sparsity and performance constraints.

4.3. Synthesis of Consensus Networks

Based on the consensus model as described above, we now state the problem of consensus design explicitly. We will utilize the fact, that due to the formulation as a feedback problem, the problem of topology design can be related to the well known problem of feedback controller design. This section will derive a problem formulation to design graphs that have an optimal performance in terms of the $\mathcal{H}_2$-norm.

We will use the formulation of the $\mathcal{H}_2$-norm of a dynamical system as given in the appendix (Theorem B.2). For the edge agreement problem, the solution of the Lyapunov equation can be written by inspection as (Zelazo and Mesbahi, 2011a)

$$X(\mathcal{G}) = \frac{1}{2} \left((R_{(\mathcal{T},\mathcal{C})} R_{(\mathcal{T},\mathcal{C})}^T)^{-1} + L_e(\mathcal{T}) \right). \tag{4.9}$$

The solution is a function of the underlying graph $\mathcal{G}$, and in particular on the choice of spanning tree and cycles. This can be used to characterize the $\mathcal{H}_2$ performance of the edge agreement problem.

Theorem 4.1 (Zelazo and Mesbahi (2011a)). *The $\mathcal{H}_2$ performance of edge agreement problem (4.4) is*

$$\|\Sigma_e(\mathcal{G})\|_2^2 = \text{Tr}[X(\mathcal{G})] = \frac{1}{2}\text{Tr}\left[(R_{(\mathcal{T},\mathcal{C})} R_{(\mathcal{T},\mathcal{C})}^T)^{-1}\right] + (n-1). \tag{4.10}$$

A direct corollary of Theorem 4.1 provides an upper and lower bound for the performance of the edge agreement problem. Of importance for the topology design is the fact that the performance is upper bounded by any choice of spanning tree and lower bounded by the complete graph $\mathcal{K}_n$.

Corollary 4.1 (Zelazo and Mesbahi (2011a)). *The $\mathcal{H}_2$ performance of the edge agreement problem for an arbitrary connected graph $\mathcal{G}$ is bounded from above and below as*

$$\|\Sigma_e(\mathcal{K}_n)\|_2^2 = \frac{n^2 - 1}{n} \leq \|\Sigma_e(\mathcal{G})\|_2^2 \leq \|\Sigma_e(\mathcal{T})\|_2^2 = \frac{3}{2}(n-1), \tag{4.11}$$

where $\mathcal{T} \subseteq \mathcal{G}$ is any spanning tree in the graph.

4.3.1. Problem Formulation – Consensus Design

In Zelazo et al. (2012) it was shown, that the $\mathcal{H}_2$ performance always improves with the addition of cycles. A trivial conclusion from Corollary 4.1 is that one always uses the complete graph. The complete graph also, for example, has the largest algebraic connectivity $\mu = \lambda_2(\mathcal{G})$ and thus is desirable for other performance indicators. However, as is common in real-world engineering applications, it may not be feasible to implement a consensus network with all possible communication links. In a more realistic scenario, the number of edges to be implemented is limited due to computation or communication cost. Additionally, some edges might simply not be possible to implement. With the above discussion in mind, one can formulate the following problem for the design of consensus networks as a compromise between performance and communication cost.

Problem 4.1 (Consensus Design). *Consider a consensus network over a spanning tree* $\mathcal{T} = (\mathcal{V}, \mathcal{E}_{\tau})$ *and a set of candidate edges* $\bar{\mathcal{E}}_{\tau}$*. Add k edges from the set* $\bar{\mathcal{E}}_{\tau}$ *to* $\mathcal{T}$ *that leads to the largest improvement in the performance of the edge agreement problem. That is, solve the following optimization problem:*

$$\min_{T_{(\mathcal{T},\mathcal{C})} \in \mathbb{R}^{|\mathcal{V}| \times k}} \|\Sigma_e(\mathcal{G})\|_2. \tag{4.12}$$

The decision to add a new cycle to the tree is a *binary* one; either a candidate edge is added or not added (the matrix $T_{(\mathcal{T},\mathcal{C})}$ takes values in $\{0, \pm 1\}$). This means that Problem 4.1 is a *mixed integer problem* (MIP) (Schrijver, 1986). In the following, we show how this problem can be transferred into a convex optimization problem and be solved using weighted ℓ_1-minimization.

Regardless of how many cycles are added, the results in Zelazo et al. (2012) also indicate that it may not be trivial to add a fixed number of cycles with the largest impact on the performance. We address this in the sequel through the introduction of an optimization problem aimed at solving (4.12). Using this formulation, we then show how additional criteria, such as maximization of the algebraic connectivity, can also be embedded into the problem.

4.3.2. Optimization Algorithm – Consensus Design

We first formulate a more detailed description of Problem (4.12). To begin, note that given a spanning tree $\mathcal{T} = (\mathcal{V}, \mathcal{E}_{\tau})$, all the candidate edges that can be added belong to the set $\bar{\mathcal{E}}_{\tau}$. In particular, this set will contain exactly $|\bar{\mathcal{E}}_{\tau}| = (1/2)(n-1)(n-2)$ edges. Furthermore, note that any graph $\mathcal{G} \subset \mathcal{K}_n$ can be obtained by the deletion of edges from the complete graph $\mathcal{K}_n$. By assigning a weight $w_i \in \{0, 1\}$ to each edge in $\bar{\mathcal{E}}_{\tau}$ we are able to represent any graph with spanning tree $\mathcal{T}$ via an appropriate choice of weights.

Let $T_{(\mathcal{T},\bar{\mathcal{T}})}$ be defined as in (C.3), describing all cycles that can be created from the spanning tree T. Then the essential edge Laplacian for any graph $\mathcal{G}$ such that $\mathcal{T} \subseteq \mathcal{G}$ can expressed as

$$L_e(\mathcal{T}) \left(I + T_{(\mathcal{T},\bar{\mathcal{T}})} W T_{(\mathcal{T},\bar{\mathcal{T}})}^T \right), \tag{4.13}$$

where $W = \text{diag}\{w_1, \ldots, w_{|\bar{\mathcal{E}}_{\tau}|}\}$, and $w_i = 1$ only for edges in $\mathcal{G}$.

Recall now from Theorem 4.1 that the performance of the edge agreement problem relates to the trace of the inverse of $\left(I + T_{(\mathcal{T},\bar{\mathcal{T}})} W T_{(\mathcal{T},\bar{\mathcal{T}})}^T \right)$. Therefore, Problem 4.1 can now be seen as the MIP

$$\min_{w_i} \text{Tr} \left[\left(I + T_{(\mathcal{T},\bar{\mathcal{T}})} W T_{(\mathcal{T},\bar{\mathcal{T}})}^T \right)^{-1} \right] \tag{4.14a}$$

$$\text{subject to } \sum_i w_i = k, \quad w_i \in \{0, 1\}. \tag{4.14b}$$

The constraint (4.14b) is used to specify how many edges are allowed to be used for cycle design.

The objective (4.14a) is a nonlinear function of the variables w_i. However, by introducing a new variable , the minimization problem can be converted to the mixed-integer semi-definite program. Consider the symmetric matrix $M \in \mathbb{R}^{|\mathcal{E}_T| \times |\mathcal{E}_T|}$ with $M \geq 0$, then (4.14) can be written as

$$\min_{M, w_i} \ \mathrm{Tr}\,[M] \tag{4.15a}$$

$$\text{s.t.} \quad \begin{bmatrix} M & I \\ I & I + T_{(\mathcal{T},\overline{\mathcal{T}})} W T_{(\mathcal{T},\overline{\mathcal{T}})} \end{bmatrix} \geq 0 \tag{4.15b}$$

$$\sum_i w_i = k, \quad w_i \in \{0,1\}. \tag{4.15c}$$

This can be seen by noting that

$$\mathrm{Tr}\left[\left(I + T_{(\mathcal{T},\overline{\mathcal{T}})} W T_{(\mathcal{T},\overline{\mathcal{T}})}^T \right)^{-1} \right] \leq \mu$$

is equivalent to the matrix inequality

$$\left(I + T_{(\mathcal{T},\overline{\mathcal{T}})} W T_{(\mathcal{T},\overline{\mathcal{T}})}^T \right)^{-1} \leq M,$$

where M is some positive semi-definite matrix satisfying $\mathrm{Tr}\,[M] \leq \mu$; this results from the fact that the trace operator is monotonic under matrix inequalities (Dullerud and Paganini, 2000). Applying the Schur complement (A.1) then yields the LMI in (4.15b). The objective then becomes the minimization of the trace of M, leading to the formulation in (4.15a).

With this reformulation, we can bow relax the mixed integer problem. The most common approach for relaxation of $\{0,1\}$-optimization problems is to relax the constraints on the weights and allow them to take values continuously on the interval $[0,1]$. This relaxation results in a convex formulation, however there is no guarantee that the solution will be integer, or for that matter, even sparse. This has several disadvantages: On the one hand, solutions will lead to a *weighted* agreement problem. More importantly, the constraint (4.14b) now takes a different interpretation. To guarantee sparse solutions, we reformulate the objective function (4.15a) in the following way as

$$\min_{M, w_i} \ \mathrm{Tr}\,[M] + \|w\|_0. \tag{4.16}$$

Recall that the ℓ_0-norm of the vector is a measure of its sparsity (as defined in Chapter 2). In this way, minimizing $\|w\|_0$ attempts to maximize the number of zero-elements in the vector. We now use the weighted ℓ_1-minimization to relax the non convex ℓ_0-minimization (see Chapter 2). The relaxed objective function reads then as follows

$$\min_{M, w_i} \ \mathrm{Tr}\,[M] + \sum_{i=1}^{n} m_i w_i, \tag{4.17}$$

where $m_1, m_2, \ldots, m_n$ are non-negative weights.

Substituting (4.17) for (4.15a) brings us to the complete ℓ_1 relaxation of the MIP described in (4.14),

$$\min_{M, w_i} \ \alpha \mathrm{Tr}\,[M] + (1 - \alpha) \sum_i m_i w_i \tag{4.18a}$$

$$\text{s.t.} \quad \begin{bmatrix} M & I \\ I & I + T_{(T,\overline{T})} W T_{(T,\overline{T})} \end{bmatrix} \geq 0 \tag{4.18b}$$

$$\sum_i w_i = k \tag{4.18c}$$

$$0 \leq w_i \leq 1. \tag{4.18d}$$

Here we have also introduced a weighting factor $\alpha \in [0, 1]$ as a tuning parameter for the relative emphasis on each term in the objective function.

It is well known, that by the introduction of re-weighted ℓ_1-minimization (Candès et al., 2008) the sparsity of the achieved solution can be increased and therefore, in the considered setup, an integer solution more easily be achieved. We therefore propose the following algorithm.

Algorithm 4.1 Topology design for consensus problems.

---

1. Set $h = 0$ and $m_i^{(0)}$ for $i = 1, \ldots, |\mathcal{E}_c|$.

2. Solve the minimization problem (4.18) to find the optimal solution $w_i^{(h)}$.

3. Update the weights

$$m_i^{(h+1)} = \frac{1}{w_i^{(h)} + \epsilon},$$

where $\epsilon > 0$ ensures that the inverse is always well defined.

4. Terminate if the solution is integer, otherwise set $h = h + 1$.

Remark 4.1. *In the author's experience, Algorithm 4.1 reaches an integer solution in 2-3 iterations for most cases.*

Remark 4.2. *While the ℓ_1-relaxation leads to a semi-definite program that can be solved in polynomial time, there are still limitations to the solvable problem sizes. In practice, this algorithm performs well for graphs with $|\mathcal{V}| < 100$ nodes (i.e., $|\overline{\mathcal{E}}_\tau| = O(1000)$). This is a consequence of considering all possible edges in the problem formulation.*

As discussed earlier, the first step of the algorithm, i.e., the initial choice of edge weights, greatly influences the solution. From an engineering design point of view, this is an advantage as it introduces additional degrees of freedom. Additional features a designer might want to promote when solving Problem 4.1 without including additional constraints into Algorithm 4.1 can be easily addressed.

Another advantage of the formulation presented in (4.18) is the ability to embed additional constraints or performance criteria into the problem. One of the most studied performance criteria for consensus networks is the rate of convergence of the system or the algebraic connectivity of the underlying graph.

The algebraic connectivity of the graph can be determined by solving a semi-definite program (Boyd, 1998),

$$\max_{\mu} \mu$$

$$\text{subject to } P^T L(\mathcal{G}) P \geq \mu I,$$

where $L(\mathcal{G})$ is the graph Laplacian, and P a matrix such that $\text{Im}\{P\} = \text{span}\{\mathbb{1}^\perp\}$. Note that for a connected graph, $P^T L(\mathcal{G}) P$ is the same size as the essential edge Laplacian.

Similar to the essential edge Laplacian, the graph Laplacian of any graph can be expressed using the complete graph with $\{0, 1\}$ weights on each possible edge. Given a spanning tree, the Laplacian can be expressed as

$$L(\mathcal{G}) = L(\mathcal{T}) + E(\mathcal{T}) T_{(\mathcal{T}, \mathcal{C})} W T_{(\mathcal{T}, \mathcal{C})}^T E(\mathcal{T})^T.$$

Note that the above equation is linear in the weights W. Therefore, this can be embedded into the program (4.18), leading to the following SDP,

$$\min_{M, \kappa, w_i} \quad \alpha_1 \text{Tr}[M] - \alpha_2 \kappa + \alpha_3 \sum_i m_i w_i \tag{4.19a}$$

$$\text{subject to } \begin{bmatrix} M & I \\ I & I + T_{(\mathcal{T}, \overline{\mathcal{T}})} W T_{(\mathcal{T}, \overline{\mathcal{T}})} \end{bmatrix} \geq 0 \tag{4.19b}$$

$$\kappa I - P^T E(\mathcal{T}) T_{(\mathcal{T}, \mathcal{C})} W T_{(\mathcal{T}, \mathcal{C})}^T E(\mathcal{T})^T P \leq PL(\mathcal{T}) P^T \tag{4.19c}$$

$$\sum_i w_i = k \tag{4.19d}$$

$$0 \leq w_i \leq 1. \tag{4.19e}$$

Remark 4.3. *At the cost of additional conservativeness, the objective function can be reduced to $\alpha_1 \text{Tr}[M] + \alpha_2 \sum_i m_i w_i$ and considering κ as a pre-defined constant. This may improve the numerical performance of the optimization.*

The program (4.19) can then be used in Algorithm 4.1. The resulting solution will be an attempted balance between the rate of convergence of the system and its $\mathcal{H}_2$ performance.

4.3.3. Numerical Example – Consensus Design

We now demonstrate the design procedure described in the previous section by a numerical example. A spanning tree with $|\mathcal{V}| = 30$ nodes was randomly created in Matlab, see Figure 4.4. The $\mathcal{H}_2$ performance of this graph is $\|\Sigma(\mathcal{T})\|_2^2 = 43.5$ and the algebraic connectivity is $\mu = 0.052$. For a complete graph, the $\mathcal{H}_2$ performance is $\|\Sigma(\mathcal{K}_n)\|_2^2 = 29.97$ and the algebraic connectivity $\mu = 30$. The longest cycle in this graph is determined by its diameter, $\mathbf{diam}[\mathcal{T}] = 10$. For this example, there are $|\overline{\mathcal{E}}_\mathcal{T}| = 406$ possible edges that can be added. We want to solve Problem 4.1 using Algorithm 4.1 and add 40 new edges to the graph. Algorithm 4.1 was implemented in Matlab using Yalmip (Löfberg, 2004) and SeDuMi (Sturm, 1999).

To emphasize the combinatorial difficulty of solving this problem exhaustively, note that there are $\binom{406}{40} \approx 3.6862 \times 10^{55}$ possibilities, while the presented algorithm converged after three updates of the ℓ_1-weights for all simulations.

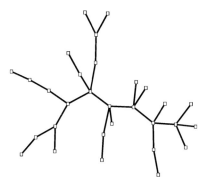

Figure 4.4.: A spanning tree on 30 nodes.

We want to show, how different initial edge-weights $m_i^{(0)}$ influence which edges are added to the tree. In fact, we emphasize that the choice of initial weights is an essential step that must be determined by the designer. In the following, we explore a variety of edge weights based on the results of Zelazo et al. (2012). A first attempt at a weighting function is to consider the inverse of the length each cycle will create; that is, the weights should be defined as

$$\text{Long cycle-length weighting: } m_i^{lc} = \mathbf{diam}[\mathcal{G}] + \mathbf{1} - \mathbf{l}(\mathbf{c_i}) \qquad (4.20)$$

where c_i is the ith column of the matrix $T_{(\mathcal{T},\mathcal{C})}$. This will place a greater emphasis on adding longer cycles in the graph.

On the other hand, it may also be desirable to encourage cycles that are short. This might arise, for example, in scenarios where sensing or communication across long distances is too costly. In this direction, we can consider weights of the form

$$\text{Short cycle-length weighting: } m_i^{sc} = l(c_i) \qquad (4.21)$$

Another weighting option is to focus on the expected number of correlated edges each cycle contains.

$$\text{Cycle-correlation weighting: } m_i^{corr} = \frac{1}{|\mathcal{E}_c|} \sum_{j \neq i} \left| \left[T_{(\mathcal{T},\mathcal{C})} \right]_{ij} \right|. \qquad (4.22)$$

Here, edges that are likely to be highly correlated with many other edges will receive a larger weight. A fourth possible weighting that attempts to balance both the cycle length and correlations is given by

$$\text{Cycle-length/correlation weighting: } m_i = \beta m_i^{lc} m_i^{corr}; \qquad (4.23)$$

the tuning parameter β can be used as a normalization factor for the weights.

Table 4.1.: Summary of $\mathcal{H}_2$ performance and $\mu(\mathcal{G})$ for each simulation example.

ℓ_1 weights	$\mathcal{H}_2$	$\mathcal{H}_2 + \mu = \lambda_2(\mathcal{G})$
m_i^{lc}	$\|\Sigma_e(\mathcal{G})\|_2^2 = 37.67$ $\mu(\mathcal{G}) = 0.28$	$\|\Sigma_e(\mathcal{G})\|_2^2 = 36.45$ $\mu(\mathcal{G}) = 0.71$
m_i^{sc}	$\|\Sigma_e(\mathcal{G})\|_2^2 = 35.27$ $\mu(\mathcal{G}) = 0.23$	$\|\Sigma_e(\mathcal{G})\|_2^2 = 36.59$ $\mu(\mathcal{G}) = 0.73$
$m_i = m_i^{lc} m_i^{corr}$	$\|\Sigma_e(\mathcal{G})\|_2^2 = 35.42$ $\mu(\mathcal{G}) = 0.08$	$\|\Sigma_e(\mathcal{G})\|_2^2 = 36.58$ $\mu(\mathcal{G}) = 0.78$
$m_i = 1$	$\|\Sigma_e(\mathcal{G})\|_2^2 = 36.73$ $\mu(\mathcal{G}) = 0.56$	$\|\Sigma_e(\mathcal{G})\|_2^2 = 36.68$ $\mu(\mathcal{G}) = 0.61$

Figure 4.5 shows the resulting graphs for different initial edge weights. In Figure 4.5(a) all initial edge weights were identical. Figure 4.5(b) uses weights favoring long cycles (4.20), Figure 4.5(c) short cycles (4.21) and Figure 4.5(d) long and correlated cycles (4.23). For m^{lc}, m^{sc} and m^{corr}, the weights were normalized to have unit norm. The combined weights were defined as in (4.23). Notice that the graph generated by the cycle length weights indeed produces a graph with longer cycles, while the other weighting options tend to favor shorter edge-disjoint cycles. For identical weights on each edge the resulting graph seems, at least qualitatively, to be in between the cycle-length and cycle-correlation results (see Figure 4.5(a)). The resulting performance for each case is given in Table 4.1.

Figure 4.6 shows the resulting graphs for the same initial edge weights, when the optimization algorithm additionally tries to maximize the algebraic connectivity. It is interesting to observe that including the connectivity constraint tends to favor longer cycles, as demonstrated by comparing the graphs in Figure 4.5 and 4.6. This reveals a subtle result relating, at least through numerical simulation, a correlation between the algebraic connectivity and the length of the cycles. It is also interesting to observe how the addition of the connectivity constraint hardly affects the attainable $\mathcal{H}_2$-performance, while the algebraic connectivity was significantly increased. The resulting performance and algebraic connectivity are also summarized in Table 4.1.

The simulation examples illustrate both the effectiveness of the design procedure and the importance of the initial edge weight selection. The variation of the performance using different edge weights are not dramatic, but the resulting graphs differ greatly. This can be considered an advantage when additional constraints on cycle length must be considered in the design process. In fact, in Zelazo et al. (2012) it was shown, that the cycle length is connected to the $\mathcal{H}_2$-performance of the consensus system.

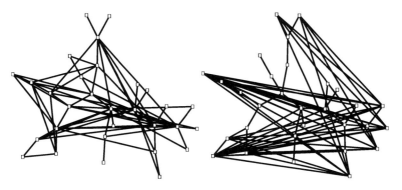

(a) Identical initial edge weights on each edge.

(b) Initial weights favoring long cycles.

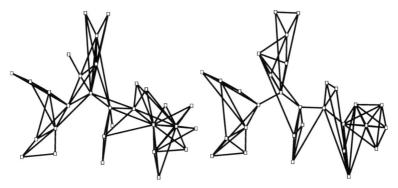

(c) Initial weights favoring short cycles.

(d) Initial weights favoring long and correlated cycles.

Figure 4.5.: Graphs for different initial edge weights.

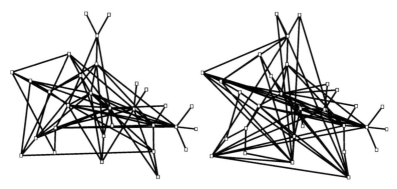

(a) Identical initial edge weights on each edge.

(b) Initial weights favoring long cycles.

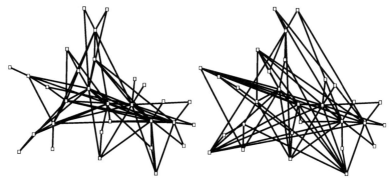

(c) Initial weights favoring short cycles.

(d) Initial weights favoring long and correlated cycles.

Figure 4.6.: Graphs for different initial edge weights with additional connectivity maximization.

4.4. Synthesis of Sparse Relative Sensing Networks

We will now state some facts for homogeneous and heterogeneous RSNs as described in the previous section.

Theorem 4.2 (Zelazo and Mesbahi (2011b)). *The $\mathcal{H}_\infty$ norm of a homogeneous RSN is given as*

$$\|T_{hom}^{w\mapsto\mathcal{G}}\|_\infty = \|WE(K_g)\|\|H\|_\infty.$$

Theorem 4.2 states that the overall $\mathcal{L}_2$-gain of the system is proportional to the matrix 2-norm of the weighted incidence matrix. For heterogeneous RSNs, only upper bounds can be given.

Theorem 4.3 (Zelazo and Mesbahi (2011b)). *The $\mathcal{H}_\infty$ norm of a heterogeneous RSN is bounded as*

$$\|T_{het}^{w\mapsto\mathcal{G}}\|_\infty \leq \|WE(K_g)^T Q\|$$

where $Q = \operatorname{diag}(\|H_1\|_\infty, \ldots, \|H_g\|_\infty)$.

Both, Theorem 4.2 and Theorem 4.3 show that the topology of the underlying graph is of significant influence to the performance of the relative sensing network. Furthermore, for heterogeneous agents, the dynamic difference between agents is an important factor in the performance of the overall system. The following sections focus on the synthesis of sparse relative sensing networks in the nominal and robust case.

The synthesis problem for (4.8), i.e. designing the topology of a RSN, can be formulated as follows:

Problem 4.2 ($\mathcal{H}_\infty$-optimal design of RSN (Zelazo and Mesbahi, 2011b)). *Given a network consisting of g agents which are coupled as given in (4.8), find edge weights $w_i \geq 0$, such that the $\mathcal{H}_\infty$-performance of the RSN is optimal, i.e.*

$$\min \; \|T_{het/hom}^{w\mapsto\mathcal{G}}\|_\infty$$
$$\text{subject to } \mathcal{G} \text{ is connected.}$$

As shown in Zelazo and Mesbahi (2011b), this problem can be formulated as a convex optimization problem. However, the solution to this optimization problem is in general not sparse, i.e. all weights w_i are non-zero. This is not a desired solution, since it requires the implementation of a complete graph. Instead, one searches for a graph topology, where most of the edge weights are zero and only very few are non-zero and have to be implemented. Therefore, in the following, we state the problem of sparse relative sensing network design. We will first consider the nominal case without uncertainties in the model and in a second step include uncertainties to the model.

4.4.1. Nominal Design of Relative Sensing Networks

We discuss different scenarios for the synthesis of homogeneous and heterogeneous relative sensing networks. The special focus is on the design of *sparse* RSNs, i.e.

networks that fulfill certain pre-specified properties with as few edges as possible. First, a problem formulation is derived that incorporates the sparsity requirements into the network design. Then, we use a weighted ℓ_1-minimization to relax the numerically exhaustive combinatorial exact solution of the original problem. Graphs with sparse topologies can then be found by the iterative solution of convex optimization problems.

Problem Formulation – Design of RSNs

Problem 4.3 (Sparse Topology Design for RSN). *Given a network consisting of g agents which are coupled as given in (4.8) and a predefined $\mathcal{H}_\infty$-performance γ. Find a sparse distribution of weights $w_i \geq 0$, such that the network is connected and the $\mathcal{H}_\infty$-performance is less than γ, i.e.*

$$\min_w \ \|w\|_0$$
$$\text{subject to } \|T^{w \to \mathcal{G}}_{het/hom}\|_\infty < \gamma$$
$$\mathcal{G} \text{ is connected.}$$

The problem formulation implies that the $\mathcal{H}_\infty$ performance of the relative sensing network does not exceed a predefined performance level γ. Recall, that the 0-norm of a vector is a measure of its sparsity. In this way, minimizing $\|w\|_0$ attempts to maximize the number of zero elements in the edge weight vector w and therefore minimizes the number of actually used edge weights.

As show in Zelazo and Mesbahi (2011b) the $\mathcal{H}_\infty$-constraint in Theorem 4.3 can be reformulated into an LMI

$$\begin{bmatrix} \gamma^2 I & QE(K_g)W \\ WE(K_g)^T Q & I \end{bmatrix} \geq 0. \tag{4.24}$$

The connectedness of the graph can be expressed as the following LMI (Boyd, 1998)

$$P^T E(K_g) W E(K_g)^T P > 0, \tag{4.25}$$

with P a matrix such that $\text{Im}\{P\} = \text{span}\{\mathbb{1}^\perp\}$. Combining (4.24) and (4.25), Problem 4.3 can now be reformulated as the following optimization problem

$$\min_w \ \|w\|_0 \tag{4.26a}$$
$$\text{subject to } \begin{bmatrix} \gamma^2 I & QE(K_g)W \\ WE(K_g)^T Q & I \end{bmatrix} \geq 0 \tag{4.26b}$$
$$P^T E(K_g) W E(K_g)^T P > 0 \tag{4.26c}$$
$$w_i \geq 0. \tag{4.26d}$$

If there is an additional constraint on the maximum weight on each edge, equation (4.26d) can be replaced by

$$0 \leq w_i \leq w_{i,\max}. \tag{4.26e}$$

Additionally, one is often not only interested in the connectedness of a graph, but also in the *maximization* of the connectivity of the graph. Since the $\mathcal{H}_\infty$-norm of the single agents $q_i = \|H_i^{yw}\|_\infty$ can be interpreted as node weights, maximization of the weighted algebraic connectivity of a graph can be formulated as (see Shafi et al. (2010))

$$\max_w \ \mu \tag{4.27}$$
$$\text{subject to } P^T(E(K_g)WE(K_g)^T - \mu Q)P > 0.$$

Note that this definition slightly differs from the classical definition of the connectivity and is associated with the node and edge- weighted graph Laplacian. To achieve a sparse topology while *simultaneously* maximizing the weighted connectivity of the graph, we combine the two objective functions (4.26a) and (4.27):

$$\min_w \ (1-\alpha)\|w\|_0 - \alpha\mu, \quad \alpha \in (0,1)$$
$$\text{subject to } \begin{bmatrix} \gamma^2 I & QE(K_g)W \\ WE(K_g)^TQ & I \end{bmatrix} \geq 0 \tag{4.28}$$
$$P^T(E(K_g)WE(K_g)^T - \mu Q)P > 0$$
$$w_i \geq 0.$$

The weighting factor $\alpha \in [0,1]$ is a tuning parameter for the relative emphasis on each term in the objective function.

Optimization Algorithm – Design of RSNs

The optimization problem (4.28) is still a combinatorial problem due to the sparsity constraints on the edge weights. We will now replace the 0-norm by its convex relaxation, the weighted 1-norm, as derived in Chapter 2. The resulting optimization problem is now convex and can be solved in polynomial time.

$$\min_w \ (1-\alpha)\sum_{i=1}^n m_i w_i - \alpha\mu, \quad \alpha \in [0,1]$$
$$\text{subject to } \begin{bmatrix} \gamma^2 I & QE(K_g)W \\ WE(K_g)^TQ & I \end{bmatrix} \geq 0 \tag{4.29}$$
$$P^T(E(K_g)WE(K_g)^T - \mu Q)P > 0$$
$$w_i \geq 0.$$

To solve the optimization problem in (4.29), we still need to choose the weighting m_i. A certain *a-priori* choice of ℓ_1-weights can be used to force the solution towards certain network topologies, as could also be seen in the previous section on cycle design. In the case of relative sensing networks, we can use the weights to promote certain sub-graphs (e.g. path graphs or star graphs). Assigning a large initial ℓ_1-weight to specific edges has the interpretation that those edges are not desirable, while small ℓ_1-weights make it more likely that those edges appear in the graph. Therefore, we propose optimization Algorithm 4.2 to seek the optimal weights m_i.

Algorithm 4.2 Sparse Topology Design Algorithm

---

1. Set $h = 0$ and choose $m_i^{(0)}$ for $i = 1, \ldots, |\mathcal{E}|$ and $\nu > 0$.

2. Solve the minimization problem (4.29) to find the optimal solution $w_i^{(h)}$.

3. Update the weights
$$m_i^{(h+1)} = (w_i^{(h)} + \nu)^{-1},$$
where $\nu > 0$ ensures that the inverse is always well defined.

4. Terminate on convergence and go to Step 5. Otherwise set $h = h + 1$ and go to Step 2.

5. Solve the optimization problem
$$\min_{w} \; -\mu$$
subject to constraints in (4.29)

for the fixed structure obtained in Step 4.

Remark 4.4. *Step 5 in Algorithm 4.2 is a so called polishing step to find the best possible edge weights for the distribution obtained by the optimization algorithm. Since ℓ_1-minimization tends to deliver weights with small absolute value, this objective is dropped in the polishing step.*

Algorithm 4.2 provides a computationally tractable solution to the original problem of sparse relative sensing network design proposed in Problem 4.5. The exhaustive combinatorial search of the 0-norm to achieve a sparse structure of the controller was relaxed by the computationally attractive weighted ℓ_1-minimization. In the next section, we will apply our results to an illustrative example.

Example – Design of RSNs

We consider networks with $g = 10$ agents to illustrate the previous results. Note that a network with 10 agents has $2^{45} \approx 3.5184 \times 10^{13}$ possible graphs (connected and unconnected). With a standard computer it is not possible to enlist all possible edge sets. However, the presented relaxation of the combinatorial problem can be solved in Matlab and Algorithm 4.2 was implemented using SeDuMi (Sturm, 1999) and YALMIP (Löfberg, 2004).

At first, let us consider that the network consists of homogenous agents. In this example, we show how an initial choice of ℓ_1-weights influences the topology of the graph and therefore, how a sub-graph can be promoted. Here, we wanted to promote a path graph as a sub-graph. Therefore, the initial ℓ_1-weights of the edges associated with a path were chosen to $m_i = 1e^{-4}$, while other initial ℓ_1-weights were set to $m_j = 1$. The edges corresponding to the path graph are the edges 1, 3, 6, 10, 15, 21, 28, 36 and 45. The predefined $\mathcal{H}_\infty$-performance was set to $\gamma = 10$.

Figure 4.7 shows the number of non-zero edges and their corresponding weighted connectivity. As can be seen, for decreasing sparsity, the weighted connectivity is increasing. In Figure 4.9 each column corresponds to a bar in Figure 4.7 with the corresponding weighted connectivity. As can be seen, the path as a subgraph is present for all weighted connectivity levels, while for increasing weighted connectivity, additional edges are added. Note that it is not possible to design a graph with a certain number of edges by the proposed optimization algorithm. Therefore, no graphs exist with e.g. 20 edges.

In the second example, we consider the network to consists of $g = 10$ heterogeneous SISO systems (generated randomly in MATLAB) with $\mathcal{H}_\infty$-performance $\|H_i\|_\infty \in [0.17, 7.48]$. The $\mathcal{H}_\infty$-performance of the RSN was specified as $\gamma = 10$ and for varying α a tradeoff between sparsity and weighted connectivity was computed. As can be seen in Figure 4.8, for increasing sparsity of the RSN, the weighted connectivity decreases. Furthermore, compared to the tree (nine edges), the weighted connectivity increases by more than 100%, when allowing 16 edges instead, while there is almost no improvement of the weighted connectivity when allowing 37 edges instead of only 23.

Figure 4.11(a), 4.11(b), 4.11(c) show the graphs for weighted connectivity level of 0.23, 0.51 and 0.84, respectively. The numbers next to the nodes correspond to the $\mathcal{H}_\infty$-norm of the single agent and represent the node weight. The darker the edge, the higher the edge weight. Note that the color of the edges only relate within one figure and are not comparable between figures. As can be seen in Figure 4.11(a), the tree is actually a star graph where the node with the highest node weight is the center node. For increasing weighted connectivity (see Figure 4.11(b) and Figure 4.11(c)), the edges connecting the next larger nodes are added.

Using Algorithm 4.2 we are able to compute a tradeoff between sparsity and connectivity of the designed relative sensing networks for large problem sizes where an exhaustive search would fail. When no special graph structures are promoted by a corresponding choice of initial values, the algorithm favors star type topologies. Since the performance of the network is predefined and the number of edges is a free variable in the optimization problem, it is not possible to design a network with a fixed number of edges.

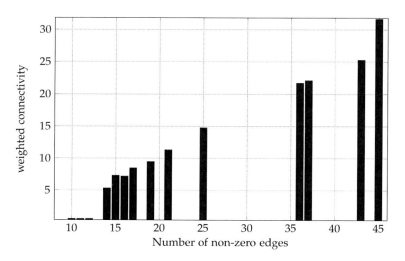

Figure 4.7.: Number of non-zero edges for homogenous RSN for increasing connectivity for $\gamma = 10$.

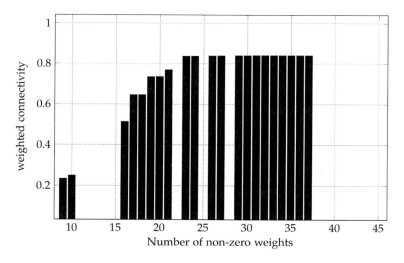

Figure 4.8.: Number of non-zero edges for heterogenous RSN for increasing connectivity for $\gamma = 10$.

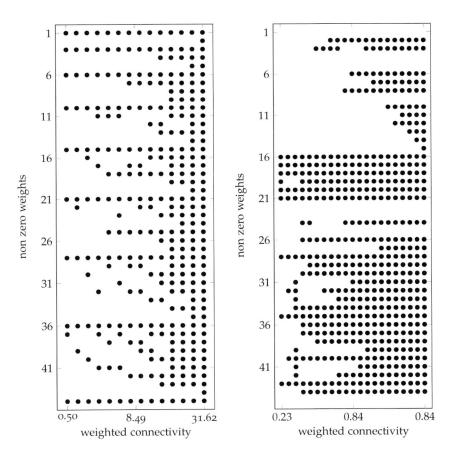

Figure 4.9.: Number of non-zero edges for heterogenous RSN for increasing connectivity for $\gamma = 10$.

Figure 4.10.: Non-zero edges for heterogenous RSN for increasing connectivity for $\gamma = 10$.

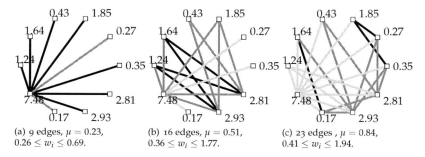

(a) 9 edges, $\mu = 0.23$,
$0.26 \leq w_i \leq 0.69$.

(b) 16 edges, $\mu = 0.51$,
$0.36 \leq w_i \leq 1.77$.

(c) 23 edges , $\mu = 0.84$,
$0.41 \leq w_i \leq 1.94$.

Figure 4.11.: Graphs with increasing number of edges and increasing connectivity for $\gamma = 10$.

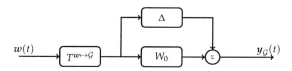

Figure 4.12.: Multiplicative uncertainty for the RSN.

4.4.2. Robust Design of Relative Sensing Networks

In addition to the nominal design of sparse relative sensing networks as presented in the previous section, we are especially interested in the *robustness* of certain topologies in face of uncertainties in the network. In the following, we assume that the agent dynamics are known exactly, whereas for the edge weight, only a nominal weight is known. We assume that the edge weights in (4.8) are given as $W = W_0 + \Delta$, where $\Delta \in \Delta_w$ is a structured uncertainty on each edge weight. This can be considered as an output-multiplicative uncertainty. The uncertainty set is defined as

$$\Delta_w = \{\text{diag}(\delta_1, \ldots, \delta_{|\mathcal{E}|}) : \delta \in \mathbb{R}^{|\mathcal{E}|}, \|\delta\|_2 \leq 1\}. \tag{4.30}$$

For the synthesis of robust RSN, we will use the term of robust connectivity defined next.

Definition 4.1 (Robust Connectivity). *A graph is called robustly connected under the uncertainty set Δ_w , if and only if the graph remains connected for all $\Delta \in \Delta_w$.*

In the next section, we use Theorem 4.3 to synthesize *robust* RSNs. We consider uncertainties in the network topology in addition to the design criteria considered in the previous section. The resulting optimization problem is now an uncertain optimization problem and therefore *infinite dimensional*. A robust counterpart is formulated and the numerically exhaustive combinatorial exact solution caused by the sparsity constraint is again relaxed by a weighted ℓ_1-minimization.

Problem Formulation – Design of Robust RSNs

We consider the network as given in (4.8) with edge weight uncertainties as in (4.30). Designing the topology for this network can be formulated as follows:

Problem 4.4 ($\mathcal{H}_\infty$-optimal design of RSN Zelazo and Mesbahi (2011b)). *Given a network consisting of g agents which are coupled as given in (4.8) with edge weight uncertainties as given in (4.30), find nominal edge weights $w_{0_i} \geq 0$, such that the*

$$\min_{w_{0_i} \geq 0} \max_{\|\delta\|_2 \leq 1} \quad \|T_{het}^{w \mapsto G}\|_\infty$$

subject to G is robustly connected.

As shown in Zelazo and Mesbahi (2011b), this problem can be formulated as a convex optimization problem using robust optimization techniques. Again, the presented solution is in general not sparse and all weights w_{0_i} are non-zero. Similar to the ideas in the previous chapter, we want to search for sparse graph topologies that require only the implementation of a few non-zero edges with additional emphasis on the robustness of the topology. To take this into account, we now state the *sparse* relative sensing network design problem with edge weight uncertainties:

Problem 4.5 (Sparse Topology Design for RSN). *Given a network consisting of g agents which are coupled as given in (4.8), edge weight uncertainty as given in (4.30) and a predefined $\mathcal{H}_\infty$-performance γ. Find a sparse distribution of the nominal weights $w_{0_i} \geq 0$, such that the network is connected and the $\mathcal{H}_\infty$-performance is less than γ, i.e.*

$$\min_{w_{0_i} \geq 0} \max_{\|\delta\|_2 \leq 1} \|w_0\|_0 \tag{4.31}$$

$$\text{subject to } \|T_{het}^{w \mapsto G}\|_\infty < \gamma$$
$$G \text{ is robustly connected.}$$

The problem formulation implies that the network remains connected and that the $\mathcal{H}_\infty$ performance of the relative sensing network does not exceed γ in the presence of edge weight uncertainties.

As shown in Zelazo and Mesbahi (2011b) Theorem 4.3 can be formulated as an uncertain LMI

$$\begin{bmatrix} \gamma^2 I & QE^1(W_0 + \Delta) \\ (W_0 + \Delta)E^T Q & I \end{bmatrix} \geq 0. \tag{4.32}$$

The algebraic connectivity of the graph can be expressed as the following uncertain LMI Boyd (1998)

$$P^T E(W_0 + \Delta)E^T P > 0, \tag{4.33}$$

with P a matrix such that $\text{Im}\{P\} = \text{span}\{\mathbb{1}^\perp\}$. Combining equation (4.32) and (4.33) leads to the following *robust optimization problem* that solves Problem 4.5

$$\min_{w_0} \max_{\|\delta\|_2 \leq 1} \|w_0\|_0 \tag{4.34a}$$

---

[1]To simplify notation, we use E instead of $E(K_g)$ from now on.

$$\text{subject to} \begin{bmatrix} \gamma^2 I & QE(W_0 + \Delta) \\ (W_0 + \Delta)E^T Q & I \end{bmatrix} \geq 0 \qquad (4.34\text{b})$$

$$P^T E(W_0 + \Delta)E^T P > 0 \qquad (4.34\text{c})$$

$$w_i \geq 0. \qquad (4.34\text{d})$$

The optimization problem for the maximization of the weighted connectivity can be formulated as follows (see Shafi et al. (2010))

$$\max_{w_0, \mu} \max_{\|\delta\|_2 \leq 1} \mu \qquad (4.35)$$

$$\text{subject to } P^T(E(W_0 + \Delta)E^T - \mu Q)P > 0.$$

To achieve a sparse topology while *simultaneously* maximizing the connectivity of the graph, we combine the two objective functions (4.34a) and (4.35) into a convex sum

$$\min_{w_0, \mu} \max_{\|\delta\|_2 \leq 1} (1 - \alpha)\|w_0\|_0 - \alpha\mu, \quad \alpha \in [0, 1] \qquad (4.36\text{a})$$

$$\text{subject to} \begin{bmatrix} \gamma^2 I & QE(W_0 + \Delta) \\ (W_0 + \Delta)E^T Q & I \end{bmatrix} \geq 0 \qquad (4.36\text{b})$$

$$P^T E(W_0 + \Delta)E^T P > 0 \qquad (4.36\text{c})$$

$$w_i \geq 0. \qquad (4.36\text{d})$$

The weighting factor $\alpha \in [0, 1]$ is a tuning parameter for the relative emphasis on each term in the objective function.

The optimization problem in (4.36) cannot directly be solved as a semidefinite program, due to two reasons. The 0-norm is a non-convex objective function and the uncertainties in the constraints result in an infinite-dimensional problem. Again, we will replace the 0-norm in the objective function by the weighted ℓ_1-minimization. For the constraints, a robust counterpart (Ben-Tal et al., 2000) can be formulated. To apply these results, the constraints in (4.36) must be rewritten in the following form

$$F^j(w, \delta) = F_0^j + \sum_i^{|\mathcal{E}|} \delta_i F_i^j(w), \quad j = 1, 2$$

where

$$F_0^1 = \begin{bmatrix} \gamma^2 I & QEW_0 \\ W_0 E^T Q & I \end{bmatrix}$$

$$F_i^1 = \begin{bmatrix} 0 & QEM_i \\ M_i E^T Q & 0 \end{bmatrix}$$

$$[M_{kl}^i] = \begin{cases} 1, & k = l = i \\ 0, & \text{otherwise,} \end{cases}$$

and

$$F_0^2 = P^T(EW_0 E^T - \mu Q)P$$

$$F_i^2 = P^T e_i e_i^T P.$$

The above expressions can now be applied to the results in Ben-Tal et al. (2000) to obtain the following robust counterpart

$$\min_{w_0,\mu,T^j,S^j} \quad (1-\alpha)\|w_0\|_0 - \alpha\mu, \quad \alpha \in [0,1] \tag{4.37a}$$

$$\text{subject to} \quad \begin{bmatrix} S^j & F_1^j & \cdots & F_{|\mathcal{E}|}^j \\ F_1^j & T^j & & \\ \vdots & & \ddots & \\ F_{|\mathcal{E}|}^j & & & T^j \end{bmatrix} \geq 0, \quad j=1,2 \tag{4.37b}$$

$$S^j + T^j \leq 2F_0^j, \quad j=1,2 \tag{4.37c}$$

$$w_i \geq 0. \tag{4.37d}$$

Remark 4.5. *It is also possible to consider $\|\delta\|_\infty \leq 1$ in equation 4.30, but this would increase the number of decision variables in the semidefinite program (4.37) even more. Therefore, we follow the slightly stricter assumption of $\|\delta\|_2 \leq 1$.*

Optimization Algorithm – Design of Robust RSNs

Using the weighted ℓ_1-minimization as proposed in Chapter 2, and the robust counterpart we can now state the convex optimization problem

$$\min_{w_0,\mu,T^j,S^j} \quad (1-\alpha)\sum_{i=1}^{n} m_i w_{0_i} - \alpha\mu, \quad \alpha \in (0,1) \tag{4.38a}$$

$$\text{subject to } (4.37\text{b}) - (4.37\text{d}) \tag{4.38b}$$

Again, the open question remains on how to choose the weights m_i. With a specific choice of weights, the solution of the optimization problem can be steered towards a certain direction, i.e. certain sub-graphs can be promoted as demonstrated in the previous section. To solve (4.38), we propose Optimization Algorithm 4.3.

Remark 4.6. *Due to the auxiliary variables introduced by the robust counterpart in (4.37), the size of the problems grows very fast with the number of nodes. Even though interior-point methods offer polynomial-time algorithms, solving the optimization problem (4.37) for very large problems might lead to numerical issues.*

Example – Design of Robust RSNs

To illustrate the previous results, we design the topology of relative sensing networks with heterogenous agents. The presented algorithm was implemented in Matlab using SeDuMi Sturm (1999) and Yalmip Löfberg (2004). In a first example, we compare the results of the combinatorial search (4.37) to the results presented in

---

Algorithm 4.3 Sparse Robust Topology Design Algorithm

---

1. Set $h = 0$ and choose $m_i^{(0)}$ for $i = 1, \ldots, |\mathcal{E}|$ and $\nu > 0$.

2. Solve the minimization problem (4.38) to find the optimal solution $w_i^{(h)}$.

3. Update the weights
$$m_i^{(h+1)} = (w_i^{(h)} + \nu)^{-1},$$
   where $\nu > 0$ ensures that the inverse is always well defined.

4. Terminate on convergence and go to Step 5. Otherwise set $h = h + 1$ and go to Step 2.

5. Solve the optimization problem
$$\min_{w_0} \ -\mu$$
$$\text{subject to } (4.37b) - (4.37d)$$
   for the fixed structure obtained in Step 4.

Algorithm 4.3. We consider a relative sensing network consisting of six agents with $\mathcal{H}_\infty$-norm $\|H_i\|_\infty \in [0.62, 6.72]$. This network can have up to 15 edges and there are 26704 possibilities for nominally connected graphs. Figure 4.13 shows the result of the comparison for a performance level of $\gamma = 18$. The bars in light gray show the maximally achievable robust connectivity for a fixed number of edges given by the exhaustive combinatorial search and the blue bars show the result achieved with Algorithm 4.3. The error due to the relaxation of the problem is very small, and we are very close to the optimal solution. The combinatorial search finds graph configurations with 7 edges, while the sparsest graph delivered by Algorithm 4.3 has 11 edges. This is due to the fact, that the weighted ℓ_1-minimization tries to make the edge weights as small as possible. But, by making the individual edge weights small, more edges are needed to ensure robust connectivity.

Second, we consider an RSN with $g = 10$ heterogeneous SISO systems, randomly generated in Matlab with $\|H_i\|_\infty \in [0.17, 7.48]$. Using the optimization algorithm presented in Zelazo and Mesbahi (2011b), the minimum $\mathcal{H}_\infty$-performance of the RSN is $\gamma_{nom} = 19.2$. The graph with the optimal performance is a complete graph with 45 non-zero edge weights. Next, we allow a slightly larger $\mathcal{H}_\infty$-performance and apply Algorithm 4.3, with $\mu = 0$ and $w_{\max} = 2$ and four iteration steps for the ℓ_1-weight update. For $\gamma = 19.34$, we can reduce the graph to 34 non-zero edges, and for $\gamma = 19.45$ to 29 edges. The results are also depicted in Fig. 4.14. Darker lines correspond to larger edge weights (note that the lines are only comparable within on graph, not between different graphs). The number of non-zero edges can be reduced by 35% by only allowing a performance degradation of 1.3%.

The third example shows the tradeoff between sparsity and weighted connectivity. Seven heterogenous SISO systems were randomly generated in Matlab with

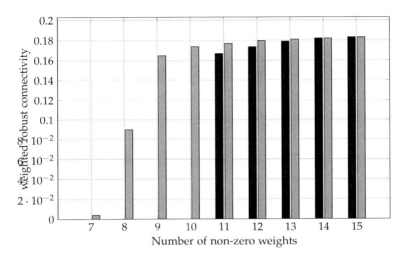

Figure 4.13.: Number of non-zero edges for heterogenous RSN for increasing connectivity for $\gamma = 18$, combinatorial approach (light gray), Algorithm 4.3 (black).

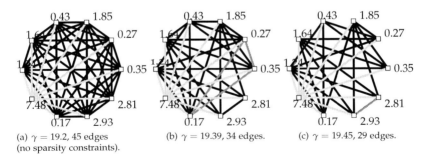

(a) $\gamma = 19.2$, 45 edges (no sparsity constraints).

(b) $\gamma = 19.39$, 34 edges.

(c) $\gamma = 19.45$, 29 edges.

Figure 4.14.: Increasing sparsity for increasing performance level γ ($w_{\max} = 2$).

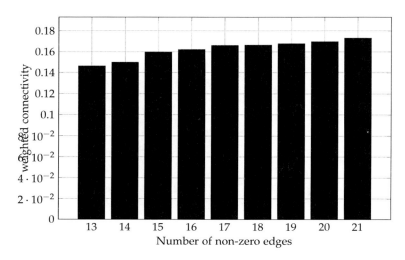

Figure 4.15.: Number of non-zero edge weights for increasing weighted connectivity for $\gamma = 10$.

$\|H_i\|_\infty \in [0.44, 3.88]$. The $\mathcal{H}_\infty$-performance of the RSN was specified as $\gamma = 10$ and for varying α a tradeoff between sparsity and weighted connectivity was computed. As can be seen in Figure 4.16, for increasing sparsity of the RSN, the weighted connectivity decreases. In Figure 4.16 each column corresponds to a bar in Figure 4.15 with the corresponding weighted connectivity. Note, that if we would solve Problem 4.5 with seven nodes as a combinatorial problem, we would have to check $\approx 1.86 \times 10^6$ possibilities of connected graph topologies. Here, we can clearly see the advantages of the weighted ℓ_1-minimization to achieve sparse structures.

4.5. Summary

This chapter focused on the design of networked dynamical systems. Special emphasis was on the topology design for open and closed loop systems. Designing networks is closely related to designing weighted and unweighted graphs and graph theory is used to study this systems. However, theoretical results exist mostly for unweighted homogenous graphs without exogenous inputs or performance outputs on the dynamical systems. Whenever node and edge weighted graphs are considered or disturbances are added to the system, almost no such results exist and synthesis has to rely on optimization techniques. Adding links to a network is a combinatorial problem – link or no link – and grows faster than polynomial with the number of edges in almost all cases. Therefore relaxation methods are needed to achieve optimization problems that are convex and solvable in polynomial time. Within the framework of weighted ℓ_1-optimization, these type of problems can be

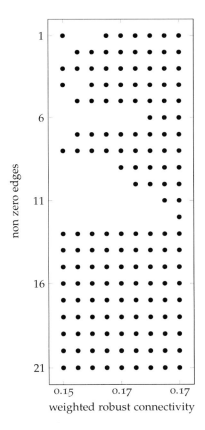

Figure 4.16.: Non-zero edge weights for increasing connectivity level. Each column depicts the non-zero edge weights for the corresponding connectivity levels in Figure 4.15.

solved numerically efficient. Due to the fact that the optimization problems always considers all possible edges, the number of optimization variables grows very fast with the number of nodes in the network. This is a drawback, when additionally uncertainties in the system are considered since the formulation of robust counterparts introduce also new variables.

We studied a canonical problem in networked control systems, namely the consensus model. Single integrator agents that interact over a network to achieve a control action. We considered the problem of adding edges to an existing tree topology such that the maximal performance improvement in terms of the $\mathcal{H}_2$-norm is achieved. This presented design algorithm turns out to be very sensitive to the weighting function used on the edges. This provides an important tuning parameter for the design of these systems. Another advantage of this representation is that additional performance parameters are easily embedded into the program. This was demonstrated by augmenting the program with an algebraic connectivity constraint. The results were demonstrated via numerical simulations.

Secondly, we considered the design of sparse relative sensing networks subject to an $\mathcal{H}_\infty$-bound on the performance. This problem is closely related to the problem of edge weight design for node and edge weighted graphs. While there exist theoretical results for $\mathcal{H}_\infty$-performance and connectivity for certain topologies of unweighted graphs, such results do not exist for weighted graphs. The problem was formulated as an optimization problem with special emphasis on the sparsity of the delivered graphs. We considered the nominal case as well as topologies with uncertainties on the edge weights. In the latter case, a robust counterpart was formulated to deal with the uncertain optimization problem. Additional performance criteria such as the maximization of the algebraic connectivity can be embedded into the resulting optimization problem. With the resulting convex optimization problem a tradeoff between sparsity and algebraic connectivity can be achieved while at the same time robustness against edge weight uncertainties is guaranteed. With an appropriate choice of initial ℓ_1-weights it is possible to influence the achieved solution. This is interesting from an engineering point of view, if special graph topologies as e.g. star or path graphs are favored.

5. ℓ_0-Gain and ℓ_1-Optimal Control

5.1. Overview

The previous two chapters focused on the design of decentralized controllers and sparse networks. We used results from compressive sensing, namely the weighted ℓ_1-minimization to achieve sparse controller matrices and sparse network topologies. Still, sparsity was only considered for static gain matrices or vectors, not for the dynamical system itself. Motivated by the success of compressive sensing in signal processing, it is reasonable to ask if there is a meaningful notion of sparsity in the context of *systems theory* and how such a concept could look like. In this chapter, a first step in this direction is made. The idea is to introduce an ℓ_0-system gain for single input/single output systems in the spirit of e.g. ℓ_2- or ℓ_∞-system gains in robust control.

In contrast to the previous two chapters were we considered continuous time systems, we will focus on discrete time systems in the following. For discrete time systems, time domain signals of the dynamical system are given in terms of infinite dimensional sequences $\{x(k)\}$ and it is possible to count the non-zero elements to evaluate the sparsity of the signal. The concept of sparsity for a continuous time signal is much more difficult to define. The ℓ_0-gain of a system is defined as the smallest ratio between the number of non-zero entries in the output signal and the input signals. Moreover, it is shown that the system gain is characterized by the number of non-zero entries of the impulse response. Therefore, a system can only be in ℓ_0 if it has finite impulse response. The second contribution of this paper is motivated by the fact that in ℓ_1-optimal control (Dahleh and Khammash, 1993; Dahleh and Diaz-Bobillo, 1995) it was observed that ℓ_1-optimal controllers produce sparse optimal closed loop impulse responses. From the view of compressive sensing and ℓ_1-minimization, this behavior is reasonable but a systems theoretic explanation of this problem seems to be not available in literature. This chapter establishes such a systems theoretic explanation of the sparse response of ℓ_1-optimal controllers by showing that the ℓ_1-optimal control problem is the best convex relaxation (in the sense of Lagrangian duality) of an appropriate ℓ_0-optimal control problem formulated with the help of the newly introduced ℓ_0-system gain. While our motivation to introduce a notion of sparsity for systems is primarily of theoretical interest, there are direct connections of the ℓ_0-gain to application areas, like the design of sparse finite impulse response filters, sparse channels and system identification via Markov parameters. In all these application areas not only the length of the impulse response is important (i.e. the last non-zero element) but also the number of non-zero elements in the impulse response, since sparse representations need less memory and can be handled computationally more efficient.

The results presented in this chapter are based on Schuler et al. (2011a).

5.2. The ℓ_0-System Gain

In this section, the concept of an ℓ_0-gain for discrete-time LTI systems is introduced. Therefore, we first define an ℓ_0-signal norm and an ℓ_0-operator norm. Moreover, a characterization of the ℓ_0-gain in terms of the number on non-zero entries in the impulse response of the system is derived.

Signal norm

Let us first introduce an ℓ_0 signal norm. We consider $x = \{x(k)\} \in \ell_0$ being a discrete-time signal with the norm

$$\|x\|_{\ell_0} := \{\text{number of } x_i | x_i \neq 0\}.$$

An alternative description which will be more useful in the given context is

$$\|x\|_{\ell_0} = \sum_{k=0}^{\infty} |\text{sign}(x(k))|.$$

Similar to the 0-norm for vectors introduced in Chapter 2.1, the ℓ_0-signal norm gives the number of non-zero elements in a discrete time signal x. Of special interest is the discrete-time impulse signal δ, with

$$\delta(k) := \begin{cases} 1 & \text{for } k = 0 \\ 0 & \text{otherwise} \end{cases}.$$

Note that the impulse signal δ fulfills $\|\delta\|_{\ell_0} = 1$.

Operator norm

We are now ready to introduce an ℓ_0-operator norm for dynamical systems. Consider a discrete-time LTI system of the following form

$$\Sigma_G : \quad \begin{bmatrix} x(k+1) \\ z(k) \end{bmatrix} = \begin{bmatrix} A & B \\ C & 0 \end{bmatrix} \begin{bmatrix} x(k) \\ w(k) \end{bmatrix} \tag{5.1a}$$

with matrices $A \in \mathbb{R}^{n \times n}$, $B \in \mathbb{R}^{n \times 1}$, and $C \in \mathbb{R}^{1 \times n}$. For the sake of simplicity, we restrict ourselves to SISO systems. It is expected that the multivariable case goes along similar lines. The impulse response g of this system is given in terms of its Markov parameters as

$$g(k) = \begin{cases} 0 & \text{for } k \leq 0 \\ CA^{k-1}B & \text{for } k > 0. \end{cases} \tag{5.2}$$

With the previously introduced ℓ_0-signal norm, we will now define the ℓ_0-gain of a system as the worst case ℓ_0 output signal for all possible input signals $w \in \ell_0$.

Definition 5.1. *The induced ℓ_0-norm (or ℓ_0-gain) of an operator $G : \ell_0 \rightarrow \ell_0$ is defined as*

$$\|G\|_{\ell_0 - ind} := \sup_{w \neq 0} \frac{\|z\|_{\ell_0}}{\|w\|_{\ell_0}}, \quad w \in \ell_0$$

with $z = Gw$.

The above definition is exactly in the spirit of systems gains known from robust control theory as e.g. the discrete time ℓ_2 or ℓ_∞-gain. A nice fact about the newly introduced ℓ_0-gain is that it is characterized by the sparsity of the impulse response of the system, as shown in the next theorem. Hence, this characterization justifies that the introduced ℓ_0-gain is a meaningful notion for sparsity.

Theorem 5.1. *The ℓ_0-gain of the system* (5.1) *is the ℓ_0-norm of its impulse response, i.e.*

$$\|G\|_{\ell_0 - ind} = \|g\|_{\ell_0}.$$

Proof. Suppose w is an arbitrary input signal with $\|w\|_{\ell_0} = N$. Then two cases can be distinguished:

1. If G has an infinite impulse response, then the ℓ_0-gain of (5.1) is infinity.

2. If G has a finite impulse response, say $\|g\|_{\ell_0} := \|G\delta\|_{\ell_0} = M$, then

$$\frac{\|z\|_{\ell_0}}{\|w\|_{\ell_0}} \leq \frac{MN}{N} = \frac{\|G\delta\|_{\ell_0}}{\|\delta\|_{\ell_0}} = \|g\|_{\ell_0}.$$

The first inequality follows from the linearity of G i.e., any output signal z is a superposition of scaled impulse responses (see Figure 5.1)

$$z(k) = \sum_{i=0}^{k} g(k-i)w(i)$$

with $g(k) = 0$, for $k > n$. Consequently, since w is arbitrary and $\|G\|_{\ell_0 - ind} \leq M$, δ is an input signal such that $\|G\|_{\ell_0 - ind} = M$.

$\square$

Theorem 5.1 gives a direct link to finite impulse response (FIR) filter, i.e. system (5.1) has a finite ℓ_0-gain if and only if it is a finite impulse response filter.

With the now defined ℓ_0-gain of a system, we want to design controllers or filters K, such that the closed loop system gain is minimal in terms of its ℓ_0-norm.

5.3. The ℓ_0-Optimal Control Problem

Consider the closed loop interconnection as depicted in Figure 5.2, where P is the generalized plant including all weighting functions and K is the controller to be designed, u is the controller output and y the measured output. The closed loop is given by $T(K) = \mathcal{F}_l(P, K)$, where $\mathcal{F}_l$ denotes the lower fractional transformation.

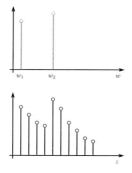

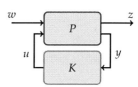

Figure 5.1.: System response as a superposition of impulse responses.

Figure 5.2.: Closed loop interconnection.

We can reformulate this closed loop in terms of its Youla Parameterization (Youla et al., 1976)

$$T(Q) = H - UQV,$$

where H, U, V are transfer functions which are derived from the state space representation of the system and Q is a free, but stable transfer function (see Appendix A.4 for details). The controller K is then given by $K = \mathcal{F}_l(J, Q)$, where J can be computed from the state space description of the generalized plant P. We can now formulate the problem of finding a transfer function Q, such that the induced norm of the closed loop system is optimal in terms of the ℓ_0-gain:

$$\mu^0 := \inf_Q \|T(Q)\|_{\ell_0-\text{ind}} = \inf_{q \in \ell_0} \|h - u * q * v\|_{\ell_0}, \tag{5.3}$$

where $*$ denotes the convolution operator and h, u, q and v are the impulse responses of H, U, Q and V, respectively. In other words, in the ℓ_0-optimal control problem, we search for an FIR filter Q such that the impulse response of the closed loop $T(Q)$ is also finite and sparse.

The optimization problem (5.3) is in general difficult to solve, since the decision variable q is wrapped between two convolution terms. Following the ideas in Khammash (2000) we will reformulate these convolution terms, such that the decision variable is only coupled to one additional variable. Let $p \in \ell_0$ be defined by $p := u * v$, and define p^k as

$$p^k := \{p(k), p(k-1), \dots, p(0), 0, \dots\}.$$

Since the convolution operation is associative and commutative we can rearrange the order of the variables

$$u * q * v = u * (q * v)$$

$$= (u * v) * q$$
$$= p * q.$$

It then follows from this equation for each time instant k

$$(u * q * v)(k) = (p * q)(k)$$
$$= \langle p^k, q \rangle,$$

where $\langle \cdot, \cdot \rangle$ denotes the standard inner product between vectors (or in this case ℓ_0-signals). With this we can reformulate the convolution terms such that the optimization variable q is only attached to the newly introduced variable p. The optimization problem (5.3) can then be rewritten as

$$\mu^0 = \inf_{q \in \ell_0} \ \|\{h(k) - \langle p^k, q \rangle\}\|_{\ell_0}. \tag{5.4}$$

Still, (5.4) is an infinite dimensional optimization problem which is non convex and difficult to solve. However, if the ℓ_0-norm of a signal x is finite, then there exists an element $x(m)$ of the sequence which is the last non-zero element of x. It then holds

$$\|x\|_{\ell_0} = \|\mathcal{P}_N(x)\|_{\ell_0}, \quad \text{for } N \geq m.$$

Consequently, if (5.4) has a solution $q^* \in l_0$ (the infimum is attained), then it is equivalent to

$$\min_{q \in \ell_0} \ \|\mathcal{P}_N(\{h(k) - \langle p^k, q \rangle\})\|_0 \tag{5.5}$$

with N sufficiently large, where $\mathcal{P}_N(x) := \{x(1), x(2), \ldots, x(N), 0, 0, \ldots\}$ is the truncation operator. With this, we have derived an equivalent formulation of the optimization problem, which is now a finite-dimensional optimization problem. Compared to the ℓ_1-optimal control problem as stated in Khammash (2000), no relaxation is necessary to transform the infinite dimensional optimization problem into a finite one, assuming that the infimum is attained. However, this problem is still far from being easy to solve. In general, N is not known in advance and can only be found by trial and error, i.e. by gradually increasing N. The second difficulty is that minimizing the ℓ_0-norm is non convex, and exhaustive combinatorial search is the only way to find the sparsest solution (Candès et al., 2006a) as discussed in Chapter 2.1. Nevertheless, in ℓ_1-optimal control, sparse impulse responses of the closed loop can be observed. Since we know that for vectors the 1-norm is the convex envelope of the 0-norm (Fazel, 2002) we want to show that an ℓ_0-optimal control problem can be constructed such that it relates in a similar way to the ℓ_1-optimal control problem. Therefore, in the next section, Lagrangian duality is applied to obtain a convex relaxation of the ℓ_0-optimal control problem.

5.4. Lagrangian Relaxation of the ℓ_0-Optimal Control Problem

In this section, we want to show that the Lagrangian relaxation of the ℓ_0-problem is the ℓ_1-problem. We do this by showing that both, the ℓ_1-problem and the

ℓ_0-problem have the same dual optimization problem. Since we know that the ℓ_1-problem is bi-dual, we can prove that the ℓ_1-problem is the bi-dual of the ℓ_0-problem and therefore its convex relaxation. At first, observe that problem (5.5) can be written similar to a standard linear program of the form

$$\min_x \ \|z\|_0 \tag{5.6a}$$

$$\text{subject to} \ \ h - Ax = z, \tag{5.6b}$$

where $h = [h(k)]$, $Ax = [\langle p^k, q \rangle]$ and $x = q$. Before we solve the problem stated in (5.6), we consider two auxiliary problems that will lead to the solution of the original problem. The idea is to augment the original optimization problem by a new term and then show that the augmented problem has the same solution as the original problem. To demonstrate this idea, we will first consider a standard ℓ_1-minimization problem. Afterwards we will again consider the previous stated ℓ_1-optimal control problem. The dual optimization problem of the ℓ_1-minimization

$$\min_x \ \|x\|_1 \tag{5.7a}$$

$$\text{subject to} \ \ Ax = b, \tag{5.7b}$$

is given by

$$\max_v \ v^T b \tag{5.8a}$$

$$\text{subject to} \ \ \|A^T v\|_\infty \leq 1. \tag{5.8b}$$

A proof of this can be found for example in Boyd and Vandenberghe (2004). Consider first the optimization problem

$$\min_x \ \|x\|_0 \tag{5.9a}$$

$$\text{subject to} \ \ Ax = b, \tag{5.9b}$$

for given $x \in \mathbb{R}^n$, $b \in \mathbb{R}^n$ and $A \in \mathbb{R}^{m \times n}$, with $m < n$ and with optimal solution x^*. In Fazel (2002), it was shown that the 1-norm is the bi-conjugate of the 0-norm and therefore its convex envelope. However, this only holds when x is restricted to a bounded domain, i.e $C = \{x \mid \|x\|_\infty < 1\}$. Otherwise, the convex envelope is identically zero. Since we do not want to restrict the domain of x, we will present an alternative approach here. Consider the optimization problem

$$\min_x \ c\|x\|_0 + \|x\|_1 \tag{5.10a}$$

$$\text{subject to} \ \ Ax = b \tag{5.10b}$$

with optimal solution $\tilde{x}^*$ where $c \in \mathbb{R}^+$ is a positive constant. This scaled optimization problem is easier to analyze and has the important property that the 0-norm of the optimal solution of (5.10) coincide with the solution of (5.9), for a sufficiently large constant c as shown in the next lemma.

Lemma 5.1. *Let x^*, $\tilde{x}^*$ be the solutions of (5.9) and (5.10), respectively. Then for sufficiently large c it holds*

$$\|x^*\|_0 = \|\tilde{x}^*\|_0.$$

Proof. The proof is by contradiction. The solution set of (5.9) and the solution set of

$$\min_x \; c\|x\|_0, \quad c > 0 \tag{5.11a}$$

$$\text{subject to } Ax = b. \tag{5.11b}$$

are identical. Remember here, that the 0-norm does not fulfill homogeneity. Let x^* be an optimal solution of (5.9) and $\tilde{x}^*$ be an optimal solution of (5.10) with $\|x^*\|_0 < \|\tilde{x}^*\|_0$. By optimality it follows

$$c\|x^*\|_0 + \|x^*\|_1 > c\|\tilde{x}^*\|_0 + \|\tilde{x}^*\|_1 \tag{5.12a}$$

$$\|x^*\|_1 > c(\underbrace{\|\tilde{x}^*\|_0 - \|x^*\|_0}_{\geq 1}) + \|\tilde{x}^*\|_1. \tag{5.12b}$$

Since x^* is finite and independent of c, i.e., x^* is a solution of (5.9) (and (5.11)) we obtain a contradiction in (5.12b) for c sufficiently large. Finally, $\|x^*\|_0 > \|\tilde{x}^*\|_0$ contradicts the assumption of optimality, since then, $\|\tilde{x}^*\|_0$ would be the optimal solution to (5.9). Therefore $\|x^*\|_0 = \|\tilde{x}^*\|_0$. $\qquad\square$

Lemma 5.1 allows us to search for sparse vectors by solving Problem (5.10) instead of Problem (5.9). Moreover, the advantage of (5.10) over (5.9) is, that its Lagrangian relaxation is not degenerated. To see this, we dualize Problem (5.10).

Lemma 5.2. *The Lagrange dual of (5.10) is given by*

$$\max_v \; v^T b$$

$$\text{subject to } \|A^T v\|_\infty \leq 1.$$

Proof. Let a_i denote the ith column of A. The Lagrangian is given by

$$L(x, v) = c\|x\|_0 + \|x\|_1 + v^T(b - Ax)$$

$$= c \sum_{i=1}^n |\text{sign}(x_i)| + \sum_{i=1}^n |x_i| + v^T b - \sum_{i=1}^n (v^T a_i) x_i$$

$$= \sum_{i=1}^n \left(c|\text{sign}(x_i)| + |x_i| - (v^T a_i) x_i \right) + v^T b.$$

The Lagrange function is

$$g(v) = \min_x L(x, v)$$

$$= \begin{cases} -\infty & \text{for } |v^T a_i| > 1 \text{ for some } i \\ v^T b & \text{for } |v^T a_i| \leq 1 \text{ for all } i. \end{cases}$$

This follows from the fact that, if $|v^T a_i| \leq 1$ then,

$$\alpha(x_i) = c|\text{sign}(x_i)| + |x_i| - (v^T a_i)x_i$$

is positive definite ($\alpha(0) = 0$, $\alpha(x_i) \geq 0$), and if $|v^T a_i| > 1$, then $\alpha(\rho) \to -\infty$ for $\rho \to \infty$. The dual problem is then given by

$$\max_v \ v^T b$$
$$\text{subject to} \ \ \|v^T A\|_\infty \leq 1.$$

$\square$

Lemma 5.2 shows that the ℓ_1-minimization problem (5.7) and the augmented ℓ_0-optimization problem (5.10) have the same dual. Therefore, the ℓ_1-minimization problem is the closest convex relaxation (in terms of Lagrangian duality) of the augmented ℓ_0-optimization problem.

After considering these two auxiliary problems, we are ready to solve the problem originally stated in (5.5) and its equivalent standard form (5.6). By introducing the new variables

$$\xi = \begin{bmatrix} z \\ x \end{bmatrix}, \ \tilde{A} = \begin{bmatrix} I & A \end{bmatrix}, \text{ and } D = \begin{bmatrix} I & 0 \end{bmatrix},$$

the minimization problem (5.6) can be rewritten as

$$\min_\xi \ \|D\xi\|_0$$
$$\text{subject to} \ \ \tilde{A}\xi = h.$$

This problem has the same form as (5.9) and therefore the same conclusions hold. For this optimization problem we will now formulate a theorem corresponding to Lemma 5.2.

Theorem 5.2. *The Lagrange relaxation of the optimization problem*

$$\min_x \ c\|z\|_0 + \|z\|_1 \tag{5.14a}$$
$$\text{subject to} \ \ h - Ax = z \tag{5.14b}$$

is given by

$$\max_v \ v^T h$$
$$\text{subject to} \ \ \|v\|_\infty \leq 1$$
$$A^T v = 0.$$

Before we can prove this theorem, we state a lemma similar to Lemma 5.1.

Lemma 5.3. *Let ξ^*, $\tilde{\xi}^*$ be the solutions of (5.6) and (5.14), respectively. Then for sufficiently large c it holds*

$$\|\xi^*\|_0 = \|\tilde{\xi}^*\|_0.$$

The proof of the lemma is similar to the proof of Lemma 5.1 and is omitted here.

Proof of Theorem 5.2. The proof goes along the lines of the proof of Lemma 5.2. With ξ, $\tilde{A}$ and D as defined before, the minimization problem can be rewritten as

$$\min_{\xi} \quad c\|D\xi\|_0 + \|D\xi\|_1$$

$$\text{subject to} \quad \tilde{A}\xi = h.$$

The Lagrangian of this problem is given by

$$L(\xi, v) = c\|D\xi\|_0 + \|D\xi\|_1 + v^T(h - \tilde{A}\xi)$$

$$= c\sum_{i=1}^{2n} d_{ii}|\text{sign}(\xi_i)| + \sum_{i=1}^{2n} d_{ii}|\xi_i| + v^T h - v^T\tilde{A}\xi$$

$$= \sum_{i=1}^{2n} d_{ii}\left(c|\text{sign}(\xi_i)| + |\xi_i|\right) + v^T h - v^T\tilde{A}\xi$$

and $d_{ii} := 0$ for $i > n$. With $\tilde{a}_i$ being the ith column of $\tilde{A}$, it follows

$$g(v) = \min_{\xi} L(\xi, v)$$

$$= \min_{\xi} \left(\sum_{i=1}^{2n} \left(d_{ii}\left(c|\text{sign}(\xi_i)| + |\xi_i|\right) - (v^T\tilde{a}_i)\xi_i\right) + v^T h\right)$$

$$= \begin{cases} -\infty & \text{for} \quad |v^T\tilde{a}_i| > d_{ii} \text{ for some } i \\ v^T h & \text{for} \quad |v^T\tilde{a}_i| \leq d_{ii} \text{ for all } i. \end{cases}$$

This is due to the fact that, if $|v^T\tilde{a}_i| \leq d_{ii}$, then

$$\alpha(x_i) = d_{ii}\left(c|\text{sign}(x_i)| + |x_i|\right) - (v^T\tilde{a}_i)x_i$$

is positive definite ($\alpha(0) = 0$, $\alpha(x_i) \geq 0$), and if $|v^T\tilde{a}_i| > d_{ii}$, then $\alpha(\rho) \to -\infty$ for $\rho \to \infty$.

The dual problem can now be written as

$$\max_{v} \quad v^T h$$

$$\text{subject to} \quad |v^T\tilde{a}_i| \leq d_{ii}$$

with $d_{ii} = 1$ for $i = 1 \ldots n$ and $d_{ii} = 0$ for $i = (n+1) \ldots 2n$. This can be rewritten as

$$\max_{v} \quad v^T h$$

$$\text{subject to} \quad \|v^T\|_\infty \leq 1$$

$$A^T v = 0.$$

$\square$

So far, we have introduced an augmented ℓ_0-problem that has the same optimal solution as the originalyl considered ℓ_0-problem. For this augmented problem, the Lagrange dual which is not degenerated has been computed. In the next section, we summarize the results obtained so far and establish the connection between the ℓ_0- and ℓ_1-optimal control problem.

5.5. The ℓ_1-Optimal Control Problem

For an LTI operator or transfer matrix, the ℓ_∞-induced norm is the ℓ_1-norm of its impulse response matrix (Zhou et al., 1996). The ℓ_∞-gain between exogenous input w and performance output z of a system, which is the ℓ_∞-induced norm of the system operator $T : \ell_\infty^q \to \ell_\infty^p$, is defined by

$$\|T\|_{\infty-\text{ind}} := \sup_{w \neq 0} \frac{\|Tw\|_\infty}{\|w\|_\infty}, \quad w \in \ell_\infty.$$

The goal of standard ℓ_1-optimal control is to design an LTI output-feedback controller $u = Ky$ that internally stabilizes the closed loop and minimizes its ℓ_∞-gain. In mathematical terms K is the argument of the optimization problem

$$\mu^1 := \inf_K \|t\|_1, \tag{5.16}$$

where the closed loop T is represented by its impulse response t. The usefulness of the ℓ_1 approach lies in the support of intuitive performance specifications directly in time-domain. It is possible to capture requirements like bounded control error, actuator saturation, bounded slope, no overshoot, etc., all in the presence of non-vanishing disturbance signals. Applications range from nano scale like atomic force microscopy (Rieber et al., 2005) to control of large wind turbines (Schuler et al., 2010b, 2013b). An in-depth treatment of the fundamentals on ℓ_1-control and of its motivation can be found in Dahleh and Diaz-Bobillo (1995) and Dahleh and Khammash (1993).

In Khammash (2000) the ℓ_1-optimal control problem is formulated in terms of the Youla parametrization as follows

$$\mu^1 = \inf_{q \in \ell_1} \|h - u * q * v\|_{\ell_1}.$$

This problem has the same structure as Problem (5.3), with the only difference that the minimization takes place in ℓ_1 instead of ℓ_0. Reordering the convolution terms as shown in Khammash (2000) leads to the optimization problem

$$\mu^1 = \inf_{q \in \ell_1} \|\{h(k) - \langle p^k, q \rangle\}\|_{\ell_1}, \tag{5.17}$$

which has the same structure as (5.4). In the same way as shown for the ℓ_0-optimal control problem, this problem can also be rewritten as

$$\min_x \|z\|_1 \tag{5.18a}$$

$$\text{subject to } h - Ax = z, \tag{5.18b}$$

where $h = [h(k)]$, $Ax = [\langle p^k, q \rangle]$ and $x = q$. For finite k the results of Section 5.4 apply and consequently the proposed ℓ_0-optimal control problem is related to the ℓ_1-optimal control problem by Lagrangian duality. Summarizing, this explains why ℓ_1-optimal control systems have sparse impulse responses.

In contrast to the ℓ_0-optimal control problem, Problem (5.17) is an infinite dimensional problem and cannot be transformed into an equivalent finite dimensional problem. To tackle this numerically, the scaled q-method was introduced in Khammash (2000). Therefore, the problem is first rewritten as

$$\min_{q \in \ell_1} \ \max\{\|h - r\|_1, \alpha\|q\|_1\}$$
$$\text{subject to } r = u * q * v,$$

and lower and upper bounds can be computed.

A lower bound for $\underline{\mu}_N(\alpha)$:

$$\underline{\mu}_N(\alpha) = \min_{q \in \ell_1} \ \max\{\|h - r\|_1, \alpha\|q\|_1\}$$
$$\text{subject to } \mathcal{P}_N(r) = \mathcal{P}_N(u * q * v)$$

A upper upper bound for $\bar{\mu}_N(\alpha)$:

$$\bar{\mu}_N(\alpha) = \min_{q \in \ell_1} \max\{\|h - r\|_1, \alpha\|q\|_1\}$$
$$\text{subject to } r = u * \mathcal{P}_N(q) * v$$

This allows to compute ℓ_1-optimal controllers and delivers the convex relaxation for suboptimal ℓ_0-controllers in terms of Lagrangian duality.

ℓ_1-optimal control results in controllers of large order and one can apply model reduction techniques, before the controller can be implemented (Schuler and Allgöwer, 2009; Ajala et al., 2010). However, the reduced models do in general not possess sparse impulse responses or sparse dynamic matrices. Therefore, one has to trade-off between large and sparse models and small but dense models.

5.6. Summary

In this chapter we have used the concept of sparse representations in compressive sensing to introduce a meaningful notion of sparsity in systems theory. We have extended the 0-norm for vectors to define an ℓ_0 signal and operator norm. With this, we have been able to introduce a ℓ_0-system gain for discrete-time LTI systems as a natural extension of the notion of sparsity from signals to systems. We have been able to show that the ℓ_0-system gain is characterized by the number of non-zero entries of the impulse response. This relates directly to ℓ_1-optimal control, where sparse closed loop system responses are observed. It was then shown that the ℓ_1-optimal control problem is the convex Lagrangian relaxation of the ℓ_0-optimal control problem, which is a mathematical explanation for this observation. While this work is motivated by the attempt to introduce a notion of sparsity for systems and by explaining the sparse response of ℓ_1-optimal controllers, it is expected that an ℓ_0-system gain may also lead to new approaches in systems analysis and controller design. Especially in network controlled systems ℓ_0-optimal control might be of interest since sparse signals often translate into low data rates.

6. Conclusion

6.1. Summary

Compressive sensing and sparse signal recovery are two of the most discussed topics in signal processing at the moment. They offer the opportunity to solve combinatorial optimization problems using convex approximations with high probability. In this thesis we have explored the possibilities these methods offer to solve *control problems*. In a world with increasing number of interconnected systems and high performance constraints, couplings and interactions between subsystems cannot be ignored any longer. On the contrary, these couplings offer e.g. the opportunity for networked dynamical systems to fulfill a common goal. By transmitting remote measurements in a decentralized control loop between individual subsystems, the performance of the overall system can be improved significantly. In this thesis, we have proposed concepts for decentralized controller design and provided tools for the relatively new topic of topology design for networked dynamical systems. Network and decentralized controller design are inherently combinatorial problems – link or no link between two agents or subsystems. Using methods from compressive sensing, namely the weighted ℓ_1-minimization, we have provided solutions of the afore mentioned problems that take their combinatorial nature directly into account. Therefore, topology design is not considered as a separate problem to be solved before the actual controller or link design occurs but integrated into one global design problem. Topology and dynamics are therefore designed *jointly*. The weighted ℓ_1-minimization was recognized as an important tool to deal with structure constraints in control theory and offers the opportunity to solve new controller and topology design problems.

In Chapter 3 we have discussed the problem of decentralized controller design with $\mathcal{H}_\infty$-performance constraints. By replacing the combinatorial objective function by an weighted ℓ_1-minimization, we were able to avoid an exhaustive search over all possibilities. Since standard linearizing transformations could not be applied to the non-convex performance constraints without loosing the controller structure, we adapted a system augmentation approach from static output feedback. This has allowed us to optimize controller structure and dynamics at the same time. Additionally, the system augmentation approach has the advantage that initial conditions for the iterative algorithms can easily be chosen, such that convergence to the global optimum is more likely. The presented approach for decentralized controller design has been applied to design oscillation and inter-area damping in power systems control.

Topology design for networked dynamical systems has been studied in Chapter 4. Not many results exist for the synthesis of homogenous and heterogenous networks that take the combinatorial nature of the graph into account. We were able to provide solutions for the consensus problem, where a fixed number of edges is added to a spanning tree and for the synthesis of relative sensing networks. The performance constraints for $\mathcal{H}_2$ and $\mathcal{H}_\infty$-performance have been formulated in terms of semidefinite programs, which allows for computationally efficient solutions. Due to these formulations, it is possible to add further performance objective to the optimization problem, such as e.g. maximization of the algebraic connectivity, as long as they can be formulated as *convex* problems.

In an attempt to further transfer the notion of sparsity from signal processing to systems theory, we have introduced the ℓ_0-gain for dynamical systems and the corresponding ℓ_0-optimal control problem. We could show that the ℓ_0-gain of a system is characterized by the number of non-zero elements of its impulse response. Additionally, in one of the main theoretical contributions of this theses, we could relate the ℓ_0-optimal control problem to the ℓ_1-optimal control problem and give a systems theoretic explanation for the sparse impulse responses observed in the closed loop response of ℓ_1-optimal control.

6.2. Outlook

To conclude this thesis we suggest some ideas on future questions and open research topics. We begin with concrete ideas related to the contributions of this thesis and continue with a more global perspective.

The approach for decentralized control presented in Chapter 3 is in general independent of the considered performance metric. The novelty lies in the incorporation of the topology design into the optimization problem, not in a specific type of performance constraint. Instead of $\mathcal{H}_\infty$-control considered in this thesis, it can be easily adapted to other metrics. Extensions to systems with model uncertainties, linear parameter-varying systems or systems with bounded nonlinearities come to mind. In fact, results for $\mathcal{H}_2$ are already reported in Lin et al. (2013). The dynamic output controllers posses a special structure and while we have discussed more general classes of dynamic output feedback in this thesis, there is still potential for research. To handle a more general class of dynamic output feedback controllers, the pattern operator has to be modified in such a way, that it is easier to promote large blocks of zeros by optimization.

The ideas on topology design for networked control systems discussed in Chapter 4 are first steps towards the design of the underlying graph topology of the network. Not many results exist so far, which directly consider the combinatorial nature of these problems. Even though the presented approach can handle the combinatorial nature, it is not applicable to very large systems. The problem size grows very fast with the number of nodes. In the case of uncertain edges, even more variables are introduced due to the formulation of a robust counterpart. Potential future research lies therefore in new methods to handle the model uncertainties, that introduce less additional variables. Possible directions are either sampling based robust optimization techniques (Calafiore, 2010) or exchange methods (Reemtsen, 1994). With robust formulations suitable for larger networks, uncertainties in the agent dynamics can also be included into the optimization problem in addition to topological uncertainties.

In Chapter 5, we have introduced an ℓ_0-gain for dynamical systems. The results so far hold for SISO systems, but we expect that the multivariable case is a direct extension. On a more application oriented side, future research topics lie in the area of finite impulse response filter design with minimum number of elements in the impulse response and the construction of sparse channels.

From a more global point of view, there are many topics in control theory, that could benefit from the results in compressive sensing and weighted ℓ_1-minimization. First steps have already been taken into different directions: Optimizing the control action in a model predictive control setup under sparsity constraints reduces transmission costs, since the signals can be compressed before being sent (Nagahara et al., 2013). Model and parameter identification with additional sparsity constraints lead to lean models with less parameters. In model reduction, weighted ℓ_1-minimization leads to simple reduced models with additionally reduced computational complexity (Löhning et al., 2011). Observability from sparse measurements was reported in Wakin et al. (2010) and sparse feedback design in Bhattacharyaand and Basar (2011).

This thesis relies heavily on the solution of semidefinite and linear programs to enforce sparsity. Achieving zeros by minimization is a numerically difficult problem and it is in general difficult to decide, whether an element with very small absolute value is zero or not. In addition, these small numbers can cause problems in multiplication and division. There are a number of general purpose solvers available and we achieved satisfying results with these within this thesis. Nevertheless, algorithms specialized to certain control problems may help in finding solutions reliably and reducing computation times significantly. As mentioned before, an important topic in this context is the conditioning of numerical solutions, which influences the sensitivity of achieved properties on slight perturbations of the solution, see e.g. Keel and Bhattacharyya (1997).

Together with the results of this thesis, the discussed problems will further stimulate the use of methods from compressive sensing and combinatorial optimization in control theory.

A. Auxiliary Results

This appendix collects results from literature concerning matrix inequalities, loop shifting, LFTs and parametrization of stabilizing controllers that are used throughout this thesis.

A.1. Matrix Inequalities

We give a short introduction in matrix inequalities and related results important for this thesis.

Definition A.1. *A matrix $X = X^T$ is called positive (semi-) definite if*

$$y^T X y > 0, \quad \forall y \in \mathbb{R}^n \{0\}, \quad \left(y^T X y \geq 0, \quad \forall y \in \mathbb{R}^n \{0\} \right).$$

It is called negative (semi-) definite if

$$y^T X y < 0, \quad \forall y \in \mathbb{R}^n \{0\}, \quad \left(y^T X y \leq 0, \quad \forall y \in \mathbb{R}^n \{0\} \right).$$

We denote a positive definite matrix by $X > 0$.

Most controller and network design problems in this thesis are formulated as semi definite optimization problems with linear matrix inequalities (LMIs) as constraints. For an in-depth treatment of LMIs from a control oriented point of view the interested reader is referred to Boyd et al. (1994). An LMI has the form

$$P(x) > 0, \quad \text{with} \quad P(x) := P_0 + \sum_{i=1}^{m} x_i P_i,$$

where $x \in \mathbb{R}^m$ is the variable and $P_i = P_i^T$. In control theory, LMIs are usually formulated in such a way, that the variables are *matrices*. For example, the well-known Lyapunov inequality for stability analysis of continuous LTI systems is equivalent to the LMI

$$A^T X + X A < 0$$

where $A \in \mathbb{R}^{n \times n}$ is the given dynamic matrix and $X = X^T$ is the matrix variable. LMIs with matrix variables can be easily transformed into LMIs with scalar variables and vice versa (Boyd et al., 1994). Meanwhile, there exist efficient solvers for LMI problems, such as e.g. SeDuMi (Sturm, 1999), which can be called comfortably from Matlab using Yalmip (Löfberg, 2004).

The well known *Schur Lemma* stated next is a useful tool to rewrite or linearize matrix inequalities (see e.g. Boyd et al., 1994, Section 2.1).

Lemma A.1. *Let* $Q = Q^T$, $R = R^T$. *Then*

$$\begin{bmatrix} Q & S \\ S^T & R \end{bmatrix} < 0$$
$$\Leftrightarrow \quad R < 0 \quad and \quad Q - SR^{-1}S^T < 0 \qquad (A.1)$$
$$\Leftrightarrow \quad Q < 0 \quad and \quad R - S^T Q^{-1}S < 0.$$

A.2. Loop Shifting

By the procedure of loop shifting, it is possible to eliminate the direct feedthrough term of a plant model for controller design. See e.g. Zhou et al. (1996, Section 12.3.4, Section 17.2) for details. We give the formulas for the continuous time case as used in Chapter 3, but equivalent formulas hold in discrete time as well. Consider a plant Σ_P with realization

$$\Sigma_P : \quad \begin{bmatrix} \dot{x} \\ z \\ y \end{bmatrix} = \begin{bmatrix} A & B_w & B_u \\ C_z & D_{zw} & D_{zu} \\ C_y & D_{yw} & D_{yu} \end{bmatrix} \begin{bmatrix} x \\ w \\ u \end{bmatrix}.$$

Suppose now that there exist a controller $\Sigma_{\hat{K}}$

$$\Sigma_{\hat{K}} : \quad \begin{bmatrix} \dot{x} \\ u \end{bmatrix} = \begin{bmatrix} \hat{A}_K & \hat{B}_K \\ \hat{C}_K & \hat{D}_K \end{bmatrix} \begin{bmatrix} x \\ \hat{y} \end{bmatrix}$$

for the plant $\Sigma_{\hat{P}}$

$$\Sigma_{\hat{P}} : \quad \begin{bmatrix} \dot{x} \\ z \\ \hat{y} \end{bmatrix} = \begin{bmatrix} A & B_w & B_u \\ C_z & D_{zw} & D_{zu} \\ C_y & D_{yw} & 0 \end{bmatrix} \begin{bmatrix} x \\ w \\ u \end{bmatrix},$$

which is sometimes more convenient than designing a controller for Σ_P. It can be shown that the controller Σ_K that yields the same closed loop map $w \mapsto z$ for Σ_P as $\Sigma_{\hat{K}}$ does for $\Sigma_{\hat{P}}$, is given by the transfer matrix $K(s) = (I + \hat{K}(s)D_{yu})^{-1}$ or in state-space form as

$$\Sigma_K : \quad \begin{bmatrix} \dot{x} \\ u \end{bmatrix} = \begin{bmatrix} \hat{A}_K - \hat{B}_K D_{yu} V \hat{C}_K & \hat{B}_K - \hat{B}_K D_{yu} V \hat{D}_K \\ V \hat{C}_K & V \hat{D}_K \end{bmatrix} \begin{bmatrix} x \\ y \end{bmatrix},$$

where $V := (I - \hat{D}_K D_{yu})^{-1}$. To understand the construction, note that $\Sigma_{\hat{P}}$ is obtained form Σ_P by the transformation $\hat{y} = y - D_{yu}u$. The result is obtained by plugging in this transformation into a realization of $u = \hat{K}\hat{y}$ and solving for u.

A.3. Linear Fractional Transformations

Linear fractional transformations (LFT) are a way of describing feedback interconnections between two systems as depicted in Figure A.1. These type of feedback

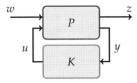

Figure A.1.: Lower linear fractional transformation.

interconnections are used in Chapter 5. The lower linear fractional transformation of two matrices P and K with appropriate partitioning is defined as

$$\mathcal{F}_l(P, K) := P_{11} + P_{12}K(I - P_{22}K)^{-1}P_{21},$$

provided the inverse $(I - P_{22}K)^{-1}$ exists.

A.4. Youla Parametrization

This section describes the Youla parametrization of all stabilizing finite-dimensional LTI output feedback controllers and all stable closed-loop maps as described in Zhou et al. (1996, Chapter 12). This parametrization was developed by Kucera (1972) and Youla et al. (1976) and is also known as the Youla-Bongiorno-Jabr-Kucera (YBJK) parametrization or the Q-parametrization. The Youla parametrization is used for discrete-time systems in this thesis (Chapter 5), and we give therefore the discrete-time discretion. Equivalent formulas hold in continuous time as well.

Lemma A.2. *Let a system Σ_P with state-space realization*

$$\Sigma_P : \begin{bmatrix} x(k+1) \\ z(k) \\ y(k) \end{bmatrix} = \begin{bmatrix} A & B_w & B_u \\ C_z & D_{zw} & D_{zu} \\ C_y & D_{yw} & D_{yu} \end{bmatrix} \begin{bmatrix} x(k) \\ w(k) \\ u(k) \end{bmatrix},$$

be given, and let F and L be such that $A + B_uF$ and $A + LC_y$ are asymptotically stable. Then all internally stabilizing finite-dimensional LTI output feedback controllers $u = Ky$ are given by

$$K = \mathcal{F}_l(J, Q)$$

where $Q \in \mathcal{RH}_\infty$, $\det(I + D_{yu}Q(\infty)) \neq 0$ and the transfer function of J is given by

$$J(z) = \left[\begin{array}{c|cc} A + B_uF + LC_y + LD_{yu}F & -L & B_u + LD_{yu} \\ \hline F & 0 & I \\ -(C_y + D_{yu}F) & I & -D_{yu} \end{array} \right].$$

Furthermore, all internally stable closed loop maps $z = P_{cl}w$ are given by

$$P_{cl}(z) = H(z) - U(z)Q(z)V(z),$$

where the transfer functions of $H(z)$, $U(z)$, and $V(z)$ are given by

$$H(z) = \left[\begin{array}{cc|c} A + B_u F & -B_u F & B_w \\ 0 & A + LC_y & B_w + LD_{yw} \\ \hline C_z + D_{zu}F & -D_{zu}F & D_{zw} \end{array} \right],$$

$$U(z) = \left[\begin{array}{c|c} A + B_u F & -B_u \\ \hline C_z + D_{zu}F & -D_{zu} \end{array} \right], \quad V(z) = \left[\begin{array}{c|c} A + LC_y & B_w + LD_{yw} \\ \hline C_y & D_{yw} \end{array} \right].$$

B. Signal and Operator Norms

The problems treated in this thesis are not only concerned with the stability of networked control systems, but also with their performance. To quantify the achieved or desired performance, norms for signals and system operators are used. These norms are defined and explained in this appendix. See for example Zhou et al. (1996), Skogestad and Postlethwaite (2005) and Dahleh and Diaz-Bobillo (1995) for further details.

B.1. Vector and Signal Norms

We denote vector norms by $\|\cdot\|_p$ and continuous time signal norms by $\|\cdot\|_{\mathcal{L}_p}$ and discrete time signal norms by $\|\cdot\|_{\ell_p}$. Notice that this distinction between vectors and discrete time signals is not necessary but is done here because of conceptual clarity.

Vector norms for $p = 0, 1, 2$ are given in Chapter 2. The $\mathcal{L}_2$-norm of a continuous time signal is defined as

$$\|v\|_{\mathcal{L}_2} = \sqrt{\int_0^\infty v(t)^T v(t) dt},$$

and measures the energy of a signal. It is defined on the Hilbert space $\mathcal{L}_2^n$ of right-sided square integrable real signals of dimension n. The ℓ_∞-norm

$$\|v\|_{\ell_\infty} = \sup_k \max_{1 \leq i \leq n} |v_i(k)|$$

measures the maximum amplitude of a sequence $v = \{v(k)\}_{k=0}^\infty$ with $v(k) = [v_1(k), \ldots, v_n(k)]^T \in \mathbb{R}^n$. The ℓ_∞-norm is defined on the space ℓ_∞^n, the Banach space of right-sided bounded real vector sequences of dimension n.

B.2. Operator Norms

Operator norms measure the "size" of a stable operator. To this end, we consider the $\mathcal{L}_p$ induced norm (or $\mathcal{L}_p$-gain)

$$\|G\|_{\mathcal{L}_p-\text{ind}} := \sup_{0 < \|w\|_{\mathcal{L}_p} < \infty} \frac{\|Gw\|_{\mathcal{L}_p}}{\|w\|_{\mathcal{L}_p}}$$

of a map $G : \mathcal{L}_p^n \mapsto \mathcal{L}_p^m$. G is said to be $\mathcal{L}_p$-stable (or just stable) if it is causal and $\|G\|_{\ell_p-\text{ind}} < \infty$. The $\mathcal{L}_p$-gain measures the worst-case amplification of the input w

in terms of the $\mathcal{L}_p$-norm. For discrete time systems, the ℓ_p-gain is similarly defined as

$$\|G\|_{\ell_p-\text{ind}} := \sup_{0<\|w\|_{\ell_p}<\infty} \frac{\|Gw\|_{\ell_p}}{\|w\|_{\ell_p}}.$$

In the case of $p = 2$, it can be shown that the $\mathcal{L}_2$-gain corresponds for a linear system G to the $\mathcal{H}_\infty$-norm

$$\|G\|_\infty = \sup_\omega\{\bar{\sigma}(G(j\omega))\},$$

where $G(s) = C(sI - A)^{-1}B + D$ is a transfer function of the dynamical system G and $\bar{\sigma}(G(j\omega))$ denotes the largest singular value of G at a fixed frequency ω. The $\mathcal{L}_2$-gain (or $\mathcal{H}_\infty$-norm) of a system can also be computed as the solution to a linear matrix inequality as formulated in the following theorem:

Lemma B.1 (Bounded Real Lemma (Gahinet and Apkarian, 1994)). *Consider a continuous-time transfer function $G(s)$ with realization $G(s) = D + C(sI - A)^{-1}B$. The following statements are equivalent*

(i) $\|D + C(sI - A)^{-1}B\|_\infty < \gamma$ and A is asymptotically stable;

(ii) there exits a symmetric positive definite solution $X > 0$ to the LMI

$$\begin{bmatrix} A^TX + XA & XB & C^T \\ B^TX & -\gamma I & D^T \\ C & D & -\gamma I \end{bmatrix} < 0. \tag{B.1}$$

In the case of $p = \infty$, the ℓ_∞-gain $\|G\|_{\ell_\infty}$ of stable LTI systems is equal to the ℓ_1-norm

$$\|G\|_1 = \max_{1\leq i\leq m} \sum_{j=1}^{n} \sum_{k=0}^{\infty} |G_{ij}(k)|$$

of the operator's impulse response $G \in \ell_1^{m\times n}$, see e.g. Zhou et al. (1996, Section 4.5). The space $\ell_1^{m\times n}$ is the Banach space of right-sided absolutely summable real matrix sequences of dimension $m \times n$.

Another system measure, not directly related to the $\mathcal{L}_p$-gains, is the $\mathcal{H}_2$-norm

$$\|G\|_2 := \sqrt{\frac{1}{2\pi} \int_{-\infty}^{\infty} \text{Tr}\left(G(j\omega) * G(j\omega)\right) d\omega},$$

defined on $\mathcal{RH}_\infty^{m\times n}$. The $\mathcal{H}_2$-norm of a system can also be computed as the solution to the following algebraic Riccati equality:

Lemma B.2 ($\mathcal{H}_2$-norm of a system (Dullerud and Paganini, 2000)). *Consider a continuous time transfer function $G(s)$ with realization $G(s) = C(sI - A)^{-1}B$ and A stable. Then we have*

$$\|G\|_2^2 = \text{Tr}(CXC^T),$$

where X is the controllability Gramian that can be obtained form the following Lyapunov equation:

$$AX + XA^T + BB^T = 0.$$

C. Graph Theory

Graph theory plays a central role in the analysis and synthesis of networked dynamical systems. This appendix gives a short introduction into graph theory with special emphasis on the tools used in this thesis (especially Chapter 4). For an in-depth treatment of graph theory, the reader is for example referred to Godsil and Royle (2009).

C.1. Graphs, Spanning Trees, and Cycles

A *graph*, denoted $\mathcal{G} = (\mathcal{V}, \mathcal{E})$, consists of a set of *nodes* $\mathcal{V} = \{v_1, \ldots, v_n\}$, and a set of *edges* $\mathcal{E}$, describing the incidence relation between pairs of nodes. In this work we deal with *undirected* graphs but at times will assign an arbitrary orientation to each edge. In this way, we denote an edge $e \in \mathcal{E}$ with the ordered pair $(v_i, v_j) \in \mathcal{V} \times \mathcal{V}$ as the *directed edge* connecting v_i to v_j; we also use the notation $v_i \sim v_j$ to denote that these two nodes are connected (or adjacent). A *path* is a sequence of distinct nodes such that consecutive nodes are adjacent to each other. If the initial and terminal node of a path are the same, it is called a *cycle*. The length of a path (cycle) is the number of edges traversed in the path sequence. For example, a *triangle* is a cycle of length 3. The *diameter* of a graph, denoted $\mathbf{diam}[\mathcal{G}]$, is the maximum distance[1] between any two nodes in a graph. A graph is *connected* if there exists a path between any pair of nodes; otherwise it is called *disconnected*. Connectedness of the graph is in this thesis assured by adding appropriate constraints to the optimization problems.

A graph $\mathcal{G}' = (\mathcal{V}', \mathcal{E}')$ is a *subgraph* of $\mathcal{G}$ if $\mathcal{V}' \subseteq \mathcal{V}$ and $\mathcal{E}' \subseteq \mathcal{E}$; equivalently, we write $\mathcal{G}' \subseteq \mathcal{G}$. A *spanning tree* of the graph $\mathcal{G} = (\mathcal{V}, \mathcal{E})$ is a connected cycle-free subgraph on the same node set, denoted $\mathcal{T} = (\mathcal{V}, \mathcal{E}_\tau) \subseteq \mathcal{G}$. Consequently, $|\mathcal{E}_\tau| = |\mathcal{V}| - 1$ for any choice of spanning tree. For a given spanning tree $\mathcal{T}$, the set $\mathcal{E}_c = \mathcal{E} \setminus \mathcal{E}_\tau$ contains all the edges in $\mathcal{G}$ that are not in $\mathcal{T}$. These edges must therefore complete the cycles of the graph, and we denote the *cycle subgraph of $\mathcal{G}$* as $\mathcal{C} = (\mathcal{V}, \mathcal{E}_c) \subseteq \mathcal{G}$. Note that the cycle subgraph depends on the choice of spanning tree, formalized by the relation $\mathcal{T} \cup \mathcal{C} = \mathcal{G}$. The *complement* of the graph $\mathcal{G} = (\mathcal{V}, \mathcal{E})$ is denoted $\overline{\mathcal{G}} = (\mathcal{V}, \overline{\mathcal{E}})$ such that $\overline{\mathcal{E}} = \{e \in \mathcal{V} \times \mathcal{V} \mid e \notin \mathcal{E}\}$. A particular graph of importance to this work is the *complete graph* on n nodes, denoted $\mathcal{K}_n$. The complete graph contains all possible edges, which for undirected graphs has cardinality $n(n-1)/2$. For any given graph $\mathcal{G}$ on n nodes, the relationship $\mathcal{G} \cup \overline{\mathcal{G}} = \mathcal{K}_n$ holds. This relation proves useful when considering the complement of a spanning tree, $\overline{\mathcal{T}}$, as it describes all possible cycles that can be created from that particular spanning tree.

---

[1] The distance between two nodes is the length of the shortest path connecting them.

Graphs also admit several useful algebraic representations. The *incidence matrix* of the graph $\mathcal{G}$, $E(\mathcal{G}) \in \mathbb{R}^{|\mathcal{V}| \times |\mathcal{E}|}$, is a $\{0, \pm 1\}$-matrix with rows and columns indexed by the vertices and edges of $\mathcal{G}$ such that $[E(\mathcal{G})]_{ik}$ has the value '+1' if node i is the initial node of edge k, ' -1' if it is the terminal node, and '0' otherwise.

The (graph) Laplacian is a symmetric and positive semi-definite matrix defined using the incidence matrix (Godsil and Royle, 2009)

$$L(\mathcal{G}) = E(\mathcal{G})E(\mathcal{G})^T. \tag{C.1}$$

Note that while the incidence matrix encodes edge directions, the Laplacian loses such information. In this work, we denote the eigenvalues of the Laplacian as

$$0 = \lambda_1(\mathcal{G}) \leq \lambda_2(\mathcal{G}) \leq \ldots \leq \lambda_{|\mathcal{V}|}(\mathcal{G}).$$

The eigenvector associated with $\lambda_1(\mathcal{G})$ is the all-ones vector, $\mathbb{1} \in \mathbb{R}^{|\mathcal{V}|}$. The algebraic connectivity of the graph $\mu = \lambda_2(\mathcal{G})$, sometimes referred to as the Fiedler eigenvalue, is strictly positive if and only if the graph is connected (Fiedler, 1973; Godsil and Royle, 2009). Similar to Shafi et al. (2010) we define the *node- and edge-weighted graph Laplacian*

$$L_w(\mathcal{G}) := Q^{-1} E(\mathcal{G}) W E(\mathcal{G})^T,$$

where Q is a positive and W a non-negative diagonal matrix representing the weights associated to the nodes and edges of the graph, respectively.

The Laplacian for the complete graph $\mathcal{K}_n$ can be expressed in terms of the Laplacian for any spanning tree $\mathcal{T}$ and the corresponding cycles $\mathcal{C} = \overline{\mathcal{T}}$ as

$$L(\mathcal{K}_n) = L(\mathcal{T}) + L(\mathcal{C}) = nI - \mathbb{1}\mathbb{1}^T. \tag{C.2}$$

Using an appropriate labeling of the edges in the graph, we can always express the incidence matrix in terms of the subgraphs $\mathcal{T}$ and $\mathcal{C}$ for a particular choice of spanning tree,

$$E(\mathcal{G}) = \begin{bmatrix} E(\mathcal{T}) & E(\mathcal{C}) \end{bmatrix}.$$

This representation aids in the interpretation of several results relating the subgraphs $\mathcal{T}$ and $\mathcal{C}$. For example, a *signed path vector* $\xi \in \mathbb{R}^{\mathcal{E}}$ is a $\{0, \pm 1\}$-vector corresponding to a path in $\mathcal{G}$, such that ξ_i takes the value '+1' ('-1') if edge $e_i \in \mathcal{E}$ is traversed positively (negatively), and '0' otherwise. Any path, therefore, can be expressed using only edges from the sub-graph $\mathcal{T}$. Observe that the length of a path can be computed from its signed path vector as $\xi^T \xi$. Furthermore, a cycle can be expressed using exactly one edge from $\mathcal{C}$, and the remaining edges from $\mathcal{T}$. This is formalized by the following result, establishing a strong connection between algebraic properties of the incidence matrix and properties of the underlying graph.

Theorem C.1 (Godsil and Royle (2009)). *For a connected graph $\mathcal{G}$, the null space of $E(\mathcal{G})$ is spanned by all the linearly independent signed path vectors corresponding to the graph cycles.*

Theorem C.1 implies that the incidence matrix corresponding to the cycle sub-graph can be expressed as a linear combination of the edges in the spanning tree. Formally, we define the matrix $T_{(\mathcal{T},\mathcal{C})} \in \mathbb{R}^{|\mathcal{V}| \times |\mathcal{E}_c|}$ as (Zelazo and Mesbahi, 2011a)

$$T_{(\mathcal{T},\mathcal{C})} = \left(E(\mathcal{T})E(\mathcal{T})^T \right)^{-1} E(\mathcal{T})^T E(\mathcal{C}), \tag{C.3}$$

satisfying

$$E(\mathcal{T})T_{(\mathcal{T},\mathcal{C})} = E(\mathcal{C}).$$

The matrix $T_{(\mathcal{T},\mathcal{C})}$, therefore, encodes information related to the cycles that can be formed from the spanning tree $\mathcal{T}$. To further aid in the following exposition, we express $T_{(\mathcal{T},\mathcal{C})}$ in terms of its columns, $T_{(\mathcal{T},\mathcal{C})} = \begin{bmatrix} c_1 & \cdots & c_{|\mathcal{E}_c|} \end{bmatrix}$, and using a slight abuse in convention, we will also refer to the ith column of $T_{(\mathcal{T},\mathcal{C})}$ as the ith cycle of the graph $\mathcal{G}$. Similarly, we will refer to the ith column of $E(\mathcal{T})$ with τ_i as the ith edge in the spanning tree. This matrix is also referred to as the *Tucker representation* of a graph, used in network optimization communities (Rockafellar, 1984). A surprising result is that the number of spanning trees that can be found in a graph, $\tau(\mathcal{G})$, can be determined as (Godsil and Royle, 2009)

$$\tau(\mathcal{G}) = \det[I + T_{(\mathcal{T},\mathcal{C})} T_{(\mathcal{T},\mathcal{C})}^T]. \tag{C.4}$$

At times, we will refer to the matrix $R_{(\mathcal{T},\mathcal{C})} = \begin{bmatrix} I & T_{(\mathcal{T},\mathcal{C})} \end{bmatrix}$. Using this notation, note that

$$E(\mathcal{G}) = E(\mathcal{T})R_{(\mathcal{T},\mathcal{C})}.$$

We explore now additional properties of the matrix $T_{(\mathcal{T},\mathcal{C})}$ and its variations. While (C.3) suggests that a matrix inverse is required to compute $T_{(\mathcal{T},\mathcal{C})}$, this is in fact unnecessary. Indeed, if the spanning tree is given, one only needs to construct a signed path vector corresponding to the edges in the tree that form a cycle with a corresponding edge in the cycle subgraph; this signed path vector will form a column of the matrix $T_{(\mathcal{T},\mathcal{C})}$. This is discussed in more detail as a consequence of the *basis theorem* in Rockafellar (1984).

C.2. The Edge Laplacian

The *edge Laplacian* is a $|\mathcal{E}| \times |\mathcal{E}|$ symmetric matrix and was introduced in Zelazo and Mesbahi (2011a) as

$$L_e(\mathcal{G}) := E(\mathcal{G})^T E(\mathcal{G}). \tag{C.5}$$

One of the results in Zelazo and Mesbahi (2011a) showed that the edge Laplacian is related to the graph Laplacian via a similarity transformation. We summarize the results here and refer the reader to Zelazo and Mesbahi (2011a) for the proof.

Theorem C.2 (Zelazo and Mesbahi (2011a)). *The graph Laplacian for a connected graph $L(\mathcal{G})$ containing cycles with an arbitrary but fixed spanning tree $\mathcal{T}$ is similar to the matrix*

$$\begin{bmatrix} L_e(\mathcal{T})R_{(\mathcal{T},\mathcal{C})}R_{(\mathcal{T},\mathcal{C})}^T & 0 \\ 0 & 0 \end{bmatrix},$$

where $L_e(\mathcal{T})$ and $R_{(\mathcal{T},\mathcal{C})} = \begin{bmatrix} I & T_{(\mathcal{T},\mathcal{C})} \end{bmatrix}$ are defined in (C.5) and (C.3), respectively.
   The edge Laplacian for a graph with cycles, $L_e(\mathcal{G})$, is similar to the matrix

$$\begin{bmatrix} L_e(\mathcal{T})R_{(\mathcal{T},\mathcal{C})}R_{(\mathcal{T},\mathcal{C})}^T & 0 \\ 0 & 0 \end{bmatrix},$$

where the block-matrix of zeros in the lower right is square with dimension equal to the number of independent cycles in the graph.
   The edge Laplacian for any graph, $L_e(\mathcal{G})$, is similar to the bordered graph Laplacian

$$\begin{bmatrix} L(\mathcal{G}) & 0 \\ 0 & 0 \end{bmatrix},$$

where the block-matrix of zeros in the lower right is square with dimension equal to the number of independent cycles in the graph minus one.

We refer to the matrix $L_e(\mathcal{T})R_{(\mathcal{T},\mathcal{C})}R_{(\mathcal{T},\mathcal{C})}^T \in \mathbb{R}^{|\mathcal{E}_T| \times |\mathcal{E}_T|}$ as the *essential edge Laplacian*. A direct consequence of Theorem C.2 is that for any connected graph the essential edge Laplacian has only positive eigenvalues, and they are precisely the non-zero eigenvalues of $L(\mathcal{G})$. Furthermore, when the underlying graph is a tree (i.e. $\mathcal{G} = \mathcal{T}$), then $L_e(\mathcal{T})R_{(\mathcal{T},\mathcal{C})}R_{(\mathcal{T},\mathcal{C})}^T = E(\mathcal{T})^T E(\mathcal{T})$ is a symmetric positive-definite matrix.

Corollary C.1 (Zelazo et al. (2012)). *The essential edge Laplacian for the complete graph $\mathcal{K}_n$ is*

$$L_e(\mathcal{T})R_{(\mathcal{T},\mathcal{C})}R_{(\mathcal{T},\mathcal{C})}^T = nI. \tag{C.6}$$

The essential edge Laplacian is the main tool used to derive an edge variant of the consensus protocol. It is important to emphasize that the similarity transformation discussed in Theorem C.2 preserves both the algebraic properties of the Laplacian, along with structural properties relating the graph to its matrix representation.

Bibliography

Ajala, O., Schuler, S., and Allgöwer, F. (2010). ℓ_∞-gain controller order reduction for discrete-time systems. In *Proc. American Control Conf. (ACC)*, 329–334.

Akyildiz, I.F., Su, W., Sankarasubramaniam, Y., and Cayirci, E. (2002). A survey on sensor networks. *IEEE Com. Magazine*, 40(8), 102–114.

Baran, T., Wei, D., and Oppenheim, A.V. (2010). Linear programming algorithms for sparse filter design. *IEEE Trans. Signal Processing*, 58(3), 1605–1617.

Barooah, P. and da Silva, N. (2006). *Distributed Optimal Estimation from Relative Measurements for Localization and Time Synchronization*. Lecture Notes in Computer Science. Springer Verlag, Berlin/Heidelberg.

Bauer, F.H., Hartman, K., How, J.P., Bristow, J., Weidow, D., and Busse, F. (1999). Enabling spacecraft formation flying through spaceborne GPS and enhanced automation technologies. In *Proc. ION-GPS Conf.*, volume 1, 369–383.

Ben-Tal, A., Ghaoui, L.E., and Nemirovski, A. (2000). Robust semidefinite programming. In H. Wolkowicz, R. Saigal, and L. Vandenberghe (eds.), *Handbook on Semidefinite Programming: Theory, Algorithms, and Applications*, chapter 6. Springer Verlag, New York, NY.

Bhattacharyaand, S. and Basar, T. (2011). Sparsity based feedback design: a new paradigm in opportunistic sensing. In *Proc. American Control Conf. (ACC)*, 3704–3709.

Blondel, V.D. and Tsitsiklis, J.N. (2000). A survey of computational complexity results in systems and control. *Automatica*, 36(9), 1249–1274.

Boyd, S.P. (1998). Convex optimization of graph Laplacian eigenvalues. In *Proc. Int. Congress of Mathematicians*, 1311–1310.

Boyd, S.P., Ghaoui, L.E., Feron, E., and Balakrishnan, V. (1994). *Linear Matrix Inequalities in System and Control Theory*. Siam Philadelphia.

Boyd, S.P. and Ghosh, A. (2006). Growing well-connected graphs. In *Proc. 45th IEEE Conf. Decision and Control (CDC)*, 6605–6611.

Boyd, S.P. and Vandenberghe, L. (2004). *Convex Optimization*. Cambridge University Press.

Briegel, B., Zelazo, D., Bürger, M., and Allgöwer, F. (2011). On the zeros of consensus networks. In *Proc. 50th IEEE Conf. on Decision and Control (CDC)*, 1890–1895.

Bristol, E.H. (1966). On a new measure of interactions for multivariable process control. *IEEE Trans. Automat. Control*, 11(1), 133–134.

Calafiore, G.C. (2010). Random convex programs. *SIAM J. on Optimization*, 20(6), 3427 – 3464.

Candès, E.J., Romberg, J.K., and Tao, T. (2006a). Robust uncertainty principles: exact signal reconstruction from highly incomplete frequency information. *IEEE Trans. Inf. Theory*, 52(2), 489–509.

Candès, E.J., Romberg, J.K., and Tao, T. (2006b). Stable signal recovery from incomplete and inaccurate measurements. *Communications on Pure and Applied Mathematics*, 59(8), 1207–1223.

Candès, E.J., Wakin, M.B., and Boyd, S.P. (2008). Enhancing sparsity by reweighted ℓ_1 minimization. *J. of Fourier Analysis and Applications*, 14(5-6), 877–905.

Cao, Y.Y., Lam, J., and Sun, Y.X. (1998). Static output feedback stabilization: An iLMI appraoch. *Automatica*, 34(12), 1641–1645.

Chow, J.H. and Cheung, K.W. (1992). A toolbox for power system dynamics and control engineering education and research. *IEEE Trans. Power Systems*, 7(4), 1559–1564.

Corazzini, T., Robertson, A., Adams, J.C., Hassibi, A., and How, J.P. (1997). GPS sensing for spacecraft formation flying 1. In *Proc. Institute of Navigation GPS-97 Conf.*

Dahleh, M.A. and Diaz-Bobillo, I.J. (1995). *Control of Uncertain Systems   A Linear Programming Approach*. Prentice Hall, Englewood Cliffs, New Jersey.

Dahleh, M.A. and Khammash, M.H. (1993). Controller design for plants with structured uncertainties. *Automatica*, 29(1), 37–56.

Dai, R. and Mesbahi, M. (2011). Optimal topology design for dynamic networks. In *Proc. 50th IEEE Conf. Decision and Control (CDC)*, 1280–1285.

Das, A., Hatano, Y., and Mesbahi, M. (2010). Agreement over noisy networks. *IET Control Theory & Applications*, 4(11), 2416.

Dinh, Q.T., Gumussoy, S., Michiels, W., and Diehl, M. (2012). Combining convex concave decompositions and linearization approaches for solving BMIs, with application to static output feedback. *IEEE Trans. Automat. Control*, 57(6), 1377–1390.

Donoho, D.L. (2006). Compressed sensing. *IEEE Trans. Inf. Theory*, 52(4), 1289–1306.

Dörfler, F., Jovanovic, M., Chertkov, M., and Bullo, F. (2013). Sparse and optimal wide-area damping control in power networks. In *Proc. American Control Conf. (ACC)*, 4295–4300.

Dullerud, G. and Paganini, F. (2000). *A Course in Robust Control Theory: A Convex Approach.* Springer Verlag, New York, NY.

Ellison, R.J., Fisher, D.A., Linger, R.C., Lispson, H.F., Longstaff, T., and Mead, N.R. (1997). Survivable network systems: An emerging discipline. Technical report, Carnegie Mellon.

Fardad, M., Lin, F., and Jovanović, M.R. (2011). Sparsity-promoting optimal control for a class of distributed systems. In *Proc. American Control Conf. (ACC)*, 2050–2055.

Fax, J.A. and Murray, R.M. (2004). Information flow and cooperative control of vehicle formations. *IEEE Trans. Automat. Control*, 49(9), 1465–1476.

Fazel, M. (2002). *Matrix Rank Minimization with Applications.* Ph.D. thesis, Stanford University.

Fazel, M., Hindi, H., and Boyd, S.P. (2003). Log-det heuristic for matrix rank minimization with applications to Hankel and Euclidean distance matrices. In *Proc. American Control Conf. (ACC)*, 2156–2162.

Fiedler, M. (1973). Algebraic connectivity of graphs. *Czechoslovak Mathematical Journal*, 23(98), 298 – 305.

Gagnepain, J.P. and Seborg, D.E. (1982). Analysis of process interactions with applications to multiloop control system design. *Ind. Eng. Chem. Process Des. Dev.*, 21, 5–11.

Gahinet, P. and Apkarian, P. (1994). A linear matrix inequality approach to $\mathcal{H}_\infty$ control. *Int. J. of Robust and Nonlinear Control*, 4, 421–448.

Ghaoui, L.E. and Oustry, F. (1997). A cone complementarity linearization algorithm for static output-feedback and related problems. *Automatic Control, IEEE*, 42(8), 1171–1176.

Giordano, P., Franchi, A., Secchi, C., and Bülthoff, H. (2011). Bilateral teleoperation of groups of UAVs with decentralized connectivity maintenance. In *Proc. Robotics: Science and Systems Conf.*, i.

Godsil, C. and Royle, G. (2009). *Algebraic Graph Theory.* Springer Verlag, New York, NY.

Grosdidier, P. and Morari, M. (1986). Interaction measures for systems under decentralized control. *Automatica*, 22(3), 309–319.

Hatano, Y. and Mesbahi, M. (2005). Agreement over random networks. *IEEE Trans. Automat. Control*, 50(11), 1867–1872.

Hatano, Y., Mesbahi, M., and Das, A. (2005). Agreement in presence of noise: pseudogradients on random geometric networks. In *Proc. 44th IEEE Conf. Decision and Control (CDC)*, 6382–6387.

He, M.J., Cai, W.J., and Wu, B.F. (2006). Control structure selection based on relative interaction decomposition. *Int. J. of Control*, 79(10), 1285–1296.

Hurley, N. and Rickard, S. (2009). Comparing measures of sparsity. *IEEE Trans. Inf. Theory*, 55(10), 4723–4741.

Iwasaki, T. and Skelton, R.E. (1994). All controllers for the general $\mathcal{H}_\infty$ control problem - LMI existence conditions and state space formulas. *Automatica*, 30, 1307–1317.

Johnston, R.D. and Barton, G.W. (1985). Structural interaction analysis. *Int. J. of Control*, 41(4), 1005–1013.

Keel, L. and Bhattacharyya, S. (1997). Robust, fragile, or optimal? *IEEE Trans. Automat. Control*, 42(8), 1098–1105.

Kerivin, H. and Mahjoub, a.R. (2005). Design of survivable networks: A survey. *Networks*, 46(1), 1–21.

Khammash, M.H. (2000). A new approach to the solution of the ℓ_1 control problem: The scaled-Q method. *IEEE Trans. Automat. Control*, 45(2), 180–187.

Khan, U.A., Member, S., Kar, S., and Moura, J.M.F. (2009). Distributed sensor localization in random environments using minimal number of anchor nodes. *IEEE Trans. Signal Processing*, 57(5), 2000–2016.

Kim, Y. and Mesbahi, M. (2005). On maximizing the second smallest eigenvalue of a state-dependent graph Laplacian. In *Proc. American Control Conf. (ACC)*, 99–103. IEEE.

Kucera, V. (1972). *Discrete Linear Control: The Polynomial Equation Approach*. Wiley, New York, NY.

Kundur, P. (1994). *Power Systems Stability and Control*. McGraw-Hill.

Leibfritz, F. (2001). An LMI-based algorithm for designing suboptimal static $\mathcal{H}_2/\mathcal{H}_\infty$ output feedback controllers. *SIAM J. Control Optim.*, 39(6), 1711–1735.

Li, P., Lam, J., Wang, Z., and Date, P. (2011). Positivity-preserving model reduction for positive systems. *Automatica*, 47(7), 1504–1511.

Lin, F., Fardad, M., and Jovanović, M.R. (2011). Algorithms for leader selection in large dynamical networks : Noise-corrupted leaders. In *Proc. 50th IEEE Conf. Decision and Control (CDC)*, 2932–2937.

Lin, F., Fardad, M., and Jovanović, M.R. (2013). Design of optimal sparse feedback gains via the alternating direction method of multipliers. *IEEE Trans. Automat. Control*, 58(9), 2426– 2431.

Löfberg, J. (2004). YALMIP: A toolbox for modeling and optimization in Matlab. In *Proc. CACSD Conf.*, 284–289.

Löhning, M., Hasenauer, J., and Allgöwer, F. (2011). Trajectory-based model reduction of nonlinear biochemical networks employing the observability normal form. In *Proc. 18th IFAC World Congress*, 10442–10447.

Luyben, W.L. (1986). Simple method for tuning SISO controllers in multivariable systems. *Ind. Eng. Chem. Process Des. Dev.*, 25, 654–660.

Mangasarian, O.L. and Pang, J.S. (1995). The extended linear complementarity problem. *SIAM J. on Matrix Analysis and Applications*, 16(2), 359–368.

Mesbahi, M. and Egerstedt, M. (2010). *Graph Theoretic Methods in Multiagent Networks*. Princeton University Press, Princeton, NJ.

Mesbahi, M. and Hadaegh, F.Y. (1999). Formation flying control of multiple spacecraft via graphs. In *IEEE Int. Conf. on Control and Application*, 1211–1216.

Mijares, G., Cole, J.D., Naugle, N.W., Preisig, H.A., and Holland, C.D. (1986). A new criterion for the pairing of control and manipulated variables. *AIChE Journal*, 32(9), 1439–1449.

Murray, R.M. (2002). *Control in an Information Rich World - Report of the Panel on Future Directions in Control, Dynamics, and Systems*. SIAM, Philadelphia.

Nabi-Abdolyousefi, M. and Mesbahi, M. (2011). Circulant networks: Controllability, observability, and linear quadratic control. *Control*, (1).

Nagahara, M., Quevedo, D.E., and Østergaard, J. (2013). Sparse packetized predictive control for networked control over erasure channels. *IEEE Trans. Automat. Control, 2014 (to appear)*.

Nedic, A. and Ozdaglar, A. (2009). Distributed subgradient methods for multiagent optimization. *IEEE Trans. Automat. Control*, 54(1), 48–61.

Niederlinski, A. (1971). A heuristic approach to design of linear multivariable interacting control systems. *Automatica*, 7(691).

Olfati-Saber, R. (2005). Distributed Kalman filter with embedded consensus filters. In *Proc. 44th IEEE Conf. Decision and Control (CDC)*, 8179–8184.

Olfati-Saber, R. and Murray, R.M. (2004). Consensus problems in networks of agents with switching topology and time-delays. *IEEE Trans. Automat. Control*, 49(9), 1520–1533.

Olfati-Saber, R. and Shamma, J. (2005). Consensus filters for sensor networks and distributed sensor fusion. *Proc. 44th IEEE Conf. Decision and Control (CDC)*, (0), 6698–6703.

Patterson, S. and Bamieh, B. (2011). Network coherence in fractal graphs. In *Proc. 50th IEEE Conf. on Decision and Control (CDC)*, 6445–6450.

Peeters, R. and Westra, R. (2004). On the identification of sparse gene regulatory networks. In *Proc. 16th Int. Symp. on Math. Theo. of Netw.*, 1–14.

Purcell, G., Kuang, D., Lichten, S., Wu, S.C., and Young, L. (1998). Autonomous formation flyer (AFF) sensor technology development. In *21st Annual AAS Guidance and Control Conf.*, 4.

Qi, X., Salapaka, M.V., Voulgaris, P.G., and Khammash, M.H. (2004). Structured optimal and robust control with multiple criteria: A convex solution. *IEEE Trans. Automat. Control*, 49(10), 1623–1640.

Rahmani, A., Ji, M., Mesbahi, M., and Egerstedt, M. (2008). Controllability of multiagent systems: from a graph-theoretic perspective. *SIAM J. on Control and Optimization*, 48(1), 162.

Reemtsen, R. (1994). Some outer approximation methods for semi-infinite optimization problems. *J. of Computational and Applied Mathematics*, 53, 87–108.

Rieber, J.M., Schitter, G., Stemmer, A., and Allgöwer, F. (2005). Experimental application of ℓ_1-optimal control in atomic force microscopy. In *Proc. 16th IFAC World Congress*, 664–669.

Rockafellar, R.T. (1984). *Network flows and monotropic optimization*. John Wiley, New York, NY.

Rotkowitz, M. and Lall, S. (2006). A characterization of convex problems in cecentralized control. *IEEE Trans. Automat. Control*, 51(2), 1984–1996.

Salgado, M.E. and Conley, A. (2004). MIMO interaction measure and controller structure selection. *Int. J. of Control*, 77(4), 367–383.

Sampei, M., Mita, T., and Nakamichi, M. (1990). An algebraic approach to $\mathcal{H}_\infty$ control problems. *Systems & Control Letters*, 14, 13–24.

Scardovi, L., Arcak, M., and Sontag, E.D. (2010). Synchronization of interconnected systems with applications to biochemical networks: An input-output approach. *IEEE Trans. Automat. Control*, 55(6), 1367–1379.

Schäfer, S. (2013). *Design of Structured Static Output Feedback Controllers for Interconnected Systems with Application to Power Systems Control*. Studienarbeit, University of Stuttgart.

Scherer, C.W. (2002). Structured finite-dimensional controller design by convex optimization. *Linear Algebra and its Applications*, 352, 639–669.

Schrijver, A. (1986). *Theory of Linear and Integer Programming*. John Wiley & Sons Ltd., West Sussex, England.

Schuler, S. and Allgöwer, F. (2009). ℓ_∞-gain model reduction for discrete-time systems via LMIs. In *Proc. American Control Conf. (ACC)*, 5701–5706.

Schuler, S., Ebenbauer, C., and Allgöwer, F. (2011a). ℓ_0-system gain and ℓ_1-optimal ontrol. In *Proc. 18th IFAC World Congress*, 9230–9235.

Schuler, S., Gruhler, M.D., Münz, U., and Allgöwer, F. (2011b). Design of structured static output feedback controllers. In *Proc. 18th IFAC World Congress*, 271–276.

Schuler, S., Li, P., Lam, J., and Allgöwer, F. (2011c). Design of structured dynamic output-feedback controllers for interconnected systems. *Int. J. of Control*, 84(12), 2081–2091.

Schuler, S., Münz, U., and Allgöwer, F. (2010a). Optimal controller structure reduction for decentralized control. In *Proc. 4th IFAC Symposium on System, Structure and Control (SSSC)*, 303–308.

Schuler, S., Münz, U., and Allgöwer, F. (2012a). Decentralized state feedback control for interconnected process systems. In *Proc. 8th IFAC Symp. Advanced Control of Chemical Processes (AdChem)*, 1–10.

Schuler, S., Münz, U., and Allgöwer, F. (2013a). Decentralized state feedback control for interconnected systems with application to power systems. *J. of Process Control*. doi:10.1016/j.jprocont.2013.10.003.

Schuler, S., Schlipf, D., and Cheng, P.W. (2013b). ℓ_1-optimal control of large wind turbines. *IEEE Trans. Control Systems Technology*, 21(4), 1079–1089.

Schuler, S., Schlipf, D., Kühn, M., and Allgöwer, F. (2010b). ℓ_1-optimal multivariable pitch control for load reduction on large wind turbines. In *Proc. Scientific Track of the European Wind Energy Conf. (EWEC)*.

Schuler, S., Zelazo, D., and Allgöwer, F. (2012b). Design of sparse relative sensing networks. In *Proc. 51th IEEE Conf. Decision and Control (CDC)*, 2749–2754.

Schuler, S., Zelazo, D., and Allgöwer, F. (2013c). Robust design of sparse relative sensing networks. In *Proc. European Control Conf. (ECC)*, 1860 – 1865.

Schuler, S., Zhou, W., Münz, U., and Allgöwer, F. (2010c). Controller structure design for decentralized control of higher order subsystems. In *Proc. 2nd IFAC Workshop on Estimation and Control of Networked Systems (NecSys)*, 269–274.

Seethalekshmi, K., Singh, S., and Srivastava, S. (2008). Wide-area protection and control: Present status and key challenges. In *Proc. 15th Nationa Power Systems Conf. (PSCE)*, 169–175.

Shafi, S.Y., Arcak, M., and Ghaoui, L.E. (2010). Designing node and edge weights of a graph to meet Laplacian eigenvalue constraints. *48th Annual Allerton Conf. on Communication, Control, and Computing*, 1016–1023.

Shah, P. and Parrilo, P.A. (2008). A partial order approach to decentralized control. In *Proc. 47th IEEE Conf. Decision and Control (CDC)*, 4351–4356.

Shu, Z. and Lam, J. (2009). An augmented system approach to static output-feedback stabilization with $\mathcal{H}_\infty$ performance for continuous-time plants. *Int. J. of Robust and Nonlinear Control*, 19(7), 768–785.

Simon, E., R-Ayerbe, P., Stoica, C., Dumur, D., and Wertz, V. (2011). LMIs-based coordinate descent method for solving BMIs in control design. In *Proc. 17th IFAC World Congress*, 10180–10186.

Skogestad, S. (2004). Control structure design for complete chemical plants. *Computers & Chemical Engineering*, 28(1-2), 219–234.

Skogestad, S. and Morari, M. (1987). Implications of large RGA elements on control performance. *Ind. Eng. Chem. Res*, 26, 2323–2330.

Skogestad, S. and Postlethwaite, I. (2005). *Multivariabe Feedback Control: Analysis and Design*. Wiley-Interscience, 2nd edition edition.

Smith, R.S. and Hadaegh, F. (2005). Control of deep-space formation-flying spacecraft; relative sensing and switched information. *J. of Guidance, Control, and Dynamics*, 28(1), 106–114.

Sturm, J.F. (1999). Using SeDuMi. *Optimization Methods and Software*, 11–12(1–4), 625–653.

Taranto, G.N., Chow, J.H., and Othman, H.A. (1994). Robust Decentralized Control Design for Damping Power System Oscillations. In *Proc. 33rd IEEE Conf. Decision and Control (CDC)*, 4080–4085.

Taranto, G.N. and Falcao, D.M. (1998). Robust delcentralised control design using genetic algorithms in power system damping control. *IEE Proc.-Gener. Trans. Distrib.*, 145(1), 1–6.

Taranto, G.N., Shiau, J.K., Chow, J.H., and Othman, H.A. (1997). Robust decentralized desing for multiple FACTS damping controllers. *IEE Proc.-Gener. Trans. Distrib.*, 144(1), 61–67.

Tonetti, S. and Murray, R.M. (2009). Limits on the Network Sensitivity Function for Multi-Agent Systems on a Graph. *Control*, 1–33.

Šiljak, D.D. (1991). *Decentralized Control of Complex Systems*. Academic Press, Bosten MA.

Wakin, M.B., Sanandaji, B.M., and Vincent, T.L. (2010). On the observability of linear systems from random, compressive measurements. *Proc. 49th IEEE Conf. on Decision and Control (CDC)*, (1), 4447–4454.

Xiao, L., Boyd, S.P., and Kim, S. (2007). Distributed average consensus with least-mean-square deviation. *J. of Parallel and Distributed Computing*, 67(1), 33–46.

Yoon, M.G., Tsu, and Mura, K. (2011). Transfer function representation of cyclic consensus systems. *Automatica*, 47(9), 1974–1982.

Youla, D.C., Jabr, H.A., and Bongiorno, J.J. (1976). Modern Wiener-Hopf design of optimal controllers - Part 2: The multivariable case. *IEEE Trans. Automat. Control*, 21(3), 319–338.

Zelazo, D. and Mesbahi, M. (2010). $\mathcal{H}_\infty$ Performance and robust topology design of relative sensing networks. In *Proc. American Control Conf. (ACC)*, 4474–4479.

Zelazo, D. and Mesbahi, M. (2011a). Edge agreement: Graph-theoretic performance bounds and passivity analysis. *IEEE Trans. Automat. Control*, 56(3), 544–555.

Zelazo, D. and Mesbahi, M. (2011b). Graph-theoretic analysis and synthesis of relative sensing networks. *IEEE Trans. Automat. Control*, 56(5), 971–982.

Zelazo, D., Schuler, S., and Allgöwer, F. (2012). Cycles and sparse design of consensus networks. In *Proc. 51th IEEE Conf. Decision and Control (CDC)*, 3808–3813.

Zelazo, D., Schuler, S., and Allgöwer, F. (2013). Performance and design of cycles in consensus networks. *Systems & Control Letters*, 62(1), 85–96.

Zhou, K., Doyle, J.o.C., and Glover, K. (1996). *Robust and Optimal Control*. Prentice Hall, Upper Saddle River, New Jersey.